AF289850

Current Topics in Microbiology 128 and Immunology

Editors

A. Clarke, Parkville/Victoria · R.W. Compans, Birmingham/Alabama · M. Cooper, Birmingham/Alabama H. Eisen, Paris · W. Goebel, Würzburg · H. Koprowski, Philadelphia · F. Melchers, Basel · M. Oldstone, La Jolla/California · R. Rott, Gießen · P.K. Vogt, Los Angeles H. Wagner, Ulm · I. Wilson, La Jolla/California

Springer-Verlag
Berlin Heidelberg NewYork Tokyo

With 12 Figures

ISBN-13: 978-3-642-71274-6   e-ISBN-13: 978-3-642-71272-2
DOI:10.1007/978-3-642-71272-2

Typesetting, printing and bookbinding:
Universitätsdruckerei H. Stürtz AG, Würzburg
2123/3130-543210

Table of Contents

Indexed in Current Contents

List of Contributors

ADA, G.L., Department of Microbiology, John Curtin School of Medical Research, Australian National University, Canberra, A.C.T. 2601, Australia

BARRETT, A.D.T., Department of Microbiology, University of Surrey, Guildford, Surrey, GU2 5XH, United Kingdom

BUTCHER, E.C., Department of Pathology, Stanford University Medical Center, Stanford, CA 94305, USA and The Veterans Administration Medical Center, Palo Alto, CA, USA

DIMMOCK, N.J., Department of Biological Sciences, University of Warwick, Conventry, CV4 7AL, United Kingdom

JONES, P.D., Department of Microbiology, John Curtin School of Medical Research, Australian National University, Canberra, A.C.T. 2601, Australia

The Immune Response to Influenza Infection

G.L. ADA and P.D. JONES

Department of Microbiology, John Curtin School of Medical Research, Australian National University, GPO Box 334, Canberra, A.C.T. 2601, Australia

Current Topics in Microbiology and Immunology, Vol. 128
© Springer-Verlag Berlin·Heidelberg 1986

1 Introduction

The purpose of this article is to review the nature of the immune response
to influenza virus both in hosts which experience natural infection, particularly
man, and in hosts which are experimentally infected, particularly mice. Until
recently, an article on this topic may well have contained only a rather superficial
account of the properties of the virion with hardly a mention of the replication
process. However, the past few years have seen great advances in two almost
separate developments. On the one hand, detailed information is now available
on the viral genome and the expression of viral proteins: there are complete
amino acid sequences and X-ray crystallographic analysis of the two surface
glycoproteins, the hemagglutinin and the neuraminidase; we have a greater
understanding of the roles of gene reassortment and mutation in the generation
of antigenic shift and drift, and of the mechanism of viral replication and assembly. On the other hand, and although there are still very significant gaps, there
has been a great increase in our knowledge of the different classes of lymphocytes, particularly T cells, and to a lesser extent of other cell types involved
in the immune response to this virus. Significant correlations in the case of
human studies and the findings from cell transfer studies in experimental systems
have made possible the allocation of particular roles to different cell types.

These two different approaches – the "virological" and the "immunological" – are now coming together. The neutralization of influenza virus by some
classes of antibody may be explicable in terms of the viral replication cycle;
evidence is beginning to accumulate that monoclonal antibodies which are very
efficient in viral neutralization react with conformational (discontinuous) determinants of the hemagglutinin; linear amino acid sequences with secondary struc-

ture may be especially effective at reacting with regulatory T cells; class I MHC antigen-restricted effector T cells may predominantly recognize membrane-bound viral protein; recognition of some internal as well as surface viral proteins by these cells explains their "apparent cross-reactivity."

Findings from each approach will increasingly impinge upon the other. Research over the next few years will result in a greater understanding of the structures seen best by both T and B cells and this may well enhance the possibility of developing better vaccines. Will an effective single peptide-based vaccine against influenza become a reality or will pronounced Ir gene effects mean that to be effective in 90% of recipients the vaccine will need to contain many viral components?

Because of this analysis of the present state of knowledge and likely future trends in the research done by many immunologists in this area, this article devotes considerable space to summarizing those aspects of the physical, chemical, and antigenic properties of the virus and its subunits and of the process of replication which seems most relevant to the recent and likely future immunological studies.

2 Chemical and Physical Properties of the Virion and Its Components

Influenza is a myxovirus, a term coined by ANDREWES et al. in 1955 to denote the affinity of the virus for mucus in the form of mucopolysaccharides and glycoproteins. Influenza viruses can be highly pleomorphic, existing as filaments and odd-shaped particles, but where well adapted, most particles are spheres with a diameter between 80 and 120 nm. When examined by negative staining, the particles show the presence of "spikes," which represent the two surface glycoproteins, the hemagglutinin (HA) and the neuraminidase (NA). The HA mediates the initial attachment of the virion to a cellular receptor and possesses a fusion capability that enables the virus envelope to integrate with an intracellular membrane, so allowing the internal components access to the cell cytoplasm. The NA is an enzyme which cleaves sialic acid residues from any oligosaccharide chain possessing that terminal sugar, including residues on both the HA and NA. The virion contains a third envelope-associated protein – the matrix (M) protein. Inside the viral envelope there are eight RNA segments and the products of five of these, the three largest protein species, the polymerase (P) proteins, supply the enzymatic machinery for viral RNA synthesis. Nucleoprotein (NP) is the fifth largest and the nonstructural proteins, NS1 and NS2, are the smallest.

Influenza viruses are divided into types A, B, and C, a classification which is based on antigenic differences between the major internal proteins, M and NP. Influenza A viruses, which are considered in this review, are described by a nomenclature which includes the host of origin, geographic origin, strain number, and year of isolation. The antigenic classification of the HA and NA are given in parenthesis, e.g., A/PR8/8/34(H1N1). There are 13 antigenic subtypes of HA (H1–H13) and 9 subtypes of NA (N1–N9). All subtypes are found in birds, but a few only in humans (H1–H3; N1–N2), swine, and horses.

2.1 The Hemagglutinin

The HA accounts for about 25% of the viral protein. It has three roles: (1) attachment of the virus to cells; (2) penetration of viral components into the cell, an early event during infection; and (3) antibodies to HA neutralize viral infectivity; antigenic variation in this molecule is mainly responsible for frequent outbreaks of influenza and for the poor control of infection by immunization. The HA molecule is present in the virion as a trimer (= "spike"). Each monomer exists as two chains, HA1 and HA2, linked by a disulfide bond. The precursor single polypeptide must be cleaved to produce HA1 and HA2 for the virus particle to be infectious. The infectivity of poorly infectious virus can be increased by trypsin treatment. HA2 is anchored in the membrane by a sequence of 25–32 hydrophobic amino acids at the C terminus. The enzyme bromelain cleaves the HA2 chain just beyond the N terminal end of the trans-membrane sequence. Some bromelain-cleaved HA preparations have been crystallized, allowing the three-dimensional structure to be ascertained by X-ray crystallography (WILEY et al. 1981; WILSON et al. 1981). Each HA molecule contains two main regions: a triple-strand coil of alpha helices and a globular region of antiparallel beta-sheets. The cell receptor-binding site and the variable antigenic determinants (see Sect. 4) are located on the globular domain. An amino sequence of ten highly conserved residues in the HA2 chain is exposed when the HA molecule is in solution at pH 5.0 and this sequence permits fusion of the molecule with the cell plasma membrane (SKEHEL et al. 1982).

Complete amino acid sequences of the HA for two H3 strains are available. Other sequences have been deduced from cDNA copies and partial sequences for 32 A strain viruses are also available. Cysteine residues and some other amino acid sequences are conserved in all cases, indicating a common ancestor end structure. The highest homology (80%) is between H3 and H5, and the lowest (25%) between H1 and H3. There is generally >90% homology between strains of a given subtype.

2.2 The Neuraminidase

The neuraminidase exists as a spike on the virion with a "head" containing four coplanar and approximately spherical subunits, a centrally attached "stalk" with a trans-membrane hydrophobic region (LAVER and VALENTINE 1969). NA is oriented in the opposite way to the HA – the trans-membrane segment is at the N terminus. It exists as a single chain and pronase treatment can release NA heads. Complete amino acid sequences of five N2 neuraminidases and partial sequences of others are known. The overall homology between N1 and N2 proteins is 39%.

The three-dimensional structure of pronase-isolated NA heads has been determined by X-ray crystallography to a 2.9-Å resolution (VARGHESE et al. 1983; COLMAN et al. 1983). The polypeptide chain is arranged in six topologically identical, four-stranded, anti-parallel beta-sheets, giving an overall appearance of propeller blades. The catalytic site appears to be a large pocket on the distal surface which is formed by some 18 residues which are conserved between subtypes.

2.3 The Nucleoprotein

The nucleoprotein has a mol. wt. of 56 kd and probably constitutes the backbone of the helical complex that is sometimes displayed by disrupted virions. Hybridization studies suggest that nucleoproteins can be divided into five groups; this grouping coincides with different host origins (BEAN 1984), which suggests that this component may be involved in determining host specificity.

2.4 The Matrix Protein

This protein is the major component in the virion and has a mol. wt. of 28 kd. The complete amino acid sequence of the protein has been deduced from base sequences and indicates that the sequences are very highly conserved. Immunization of mice with preparations of this protein gives limited protection, but the reasons for this are not clear.

2.5 Other Virion Proteins

RNA segment 8 codes for two polypeptides, NS1 and NS2. As expected, there is strong homology ($>85\%$) between the RNA from different viruses isolated over time, but substantial deletions have been detected in human field isolates.

There are three polymerase proteins: PA, PB1, and PB2, with mol. wts. of 85, 96, and 87 kd respectively, which transcribe the viral RNA. PA is thought to be responsible for virion RNA synthesis and PB1 and 2 for complementary RNA synthesis (PALESE et al. 1977). Again, nucleotide sequences suggest a high rate of conservation (complementarity $>80\%$) and amino acid sequence even higher (>95).

3 Replication and Assembly of Influenza Virus

The first step in the infection of a susceptible cell by influenza virus is the interaction of the virus with a cell receptor. The cell receptor contains sialic acid and the viral acceptor (anti-receptor) is the distal end of the HA molecule. The acceptor is a surface pocket within a 27 amino acid sequence as shown by the use of mutant viruses with altered receptor specificity. The genome of the virus particle must breach two lipid membranes – of the virus and of the cell – in order to gain entry to the cell. This is achieved by the HA molecule expressing fusion activity under appropriate conditions. If monkey kidney cells expressing the cloned HA gene are exposed to pH 5.0 buffer, the cells fuse into polykaryons (WHITE et al. 1982). The HA molecule must have been cleaved and the fusion peptide has been demonstrated to be a long stretch of highly conserved residues at the N-terminus of HA2. This region is present at the interface between monomers in the HA trimer, so the adjustment to pH 5 must result in a conformational change to expose the fusion peptide. This fusion most likely takes place after the virus particle has been endocytosed in a coated vesicle which fuses with an endosome. This provides an environment with the correct pH to allow fusion of the viral and cellular membrane which then allows the viral nucleocapsid to escape into the cytoplasm (SIMONS et al. 1982).

The first indication of the unique mechanism used by influenza virus to achieve replication came from studies showing that actinomycin D which inhibits transcription of DNA-templates also inhibits replication of the virus (BARRY et al. 1962). Alpha-amantins, known to block the action of cellular RNA polymerase II, which is involved in DNA transcription, was also shown to block influenza virus replication (MAHY et al. 1972). The role of the cellular polymerase is to provide a capped and methylated primer which is necessary for viral mRNA synthesis. The cellular mRNA primer contributes 10–15 nucleotides from the 5′ terminal end for incorporation into the viral mRNA product. Being a negative strand virus, the genomic RNA serves two functions, one transcription and the other replication. The role of the viral transcriptase is to transcribe the genomic RNA to make monocistronic mRNA (+strand), each of which specifies a single protein. The viral polymerases direct the replication process. A full-length (+) strand is made and this in turn serves as a template for the synthesis of progeny (−) strand RNA. It has been pointed out that monocistronic RNA is a convenient way to control the abundance of individual proteins (ROIZMAN 1985)

Of the eight influenza virus RNA segments, Nos. 7 and 8 each encode two proteins. Segment 7 encodes the non-glycosylated envelope protein M1 in a continous reading sequence. The smaller protein M2, also encoded in segment 7, is mainly in a reading sequence more than 700 nucleotides from the 3′ terminus of the RNA, but it also possesses 51 nucleotides coterminal with the 5′ terminus of the message for M1. Nine of the amino acids in this segment are identical in M1 and M2. M2 is not found in the virion but is expressed at the surface of infected cells. The two proteins encoded in RNA segment 8 show a similar relationship. NS1 is derived from a continuous reading sequence, whereas the RNA coding for the NS2 protein has had nucleotides 57–528 spliced out. Neither NS1 nor NS2 is found in virions and their function is not clear. mRNA molecules coding for internal proteins are synthesized earlier than mRNA coding for glycoproteins, implying some form of transcriptional regulation (HAY et al. 1977).

The assembly of the virion occurs in two stages: assembly of the nucleocapsid and inclusion of this in the viral envelope. Nucleocapsid assembly appears to take place in the nucleus, whereas the viral envelope takes shape at the cell plasma membrane. The latter process begins in the membrane of the endoplasmic reticulum in which the viral glycoproteins are first inserted and then glycosylated as they pass through the Golgi apparatus. The mechanism of tertiary folding to form NA tetramers and HA trimers is not known in detail. The

M1 protein acts as a bridge between the nucleocapsid and the viral envelope, but how this occurs and whether there is contact between the glycoproteins, intracytoplasmic segments, and nucleocapsid proteins is not clear.

The assembled virion exposes to the environment only two glycoproteins. The HA has its most important roles before replication. The NA may have functions both before and after replication. Before replication, the NA may act on nonspecific inhibitors in the extracellular fluids and possibly to release virus from mucus. After assembly of the virus, the neuraminidase acts not only on substrates in or on the infected cell, but also on the virion itself.

An aspect of increasing interest is the expression of viral proteins other than HA and NA at the surface of the infected cell. It has been shown (VIRELIZIER et al. 1977) and confirmed (YEWDELL et al. 1981) that NP is expressed at the infected cell surface and this may also occur in cells transfected with DNA encoding NP (TOWNSEND et al. 1984a). Different results have been found with matrix protein. Investigators using polyclonal antisera suggested that M was expressed at the cell surface (ADA and YAP 1977; BIDDISON et al. 1977; REISS and SCHULMAN 1980a); in contrast, workers using monoclonal antibody preparations have found only very small amounts expressed at the cell surface (HACKETT et al. 1980; YEWDELL et al. 1981). The different results may either be explained by contamination of the polyclonal antisera with anti-HA antibodies, though this has not been demonstrated and steps were taken to eliminate such a possibility (REISS and SCHULMAN 1980a); or possibly by the anti-matrix sera recognizing those amino acid sequences shared between M1 and M2. M2 protein has been found to be expressed in considerable amounts at the infected cell surface (LAMB and CHOPPIN 1983), as has NS1, a nonstructural protein (SHAW et al. 1981).

The extent to which other internal virion proteins may be expressed at the infected cell surface is not clear. It is assumed that a similar surface expression pattern occurs in infected cells in vivo.

4 Antigenic Properties of the Viral Proteins

Until recently, this was only known in detail for antibody reactions but there is now increasing interest in T-cell recognition, so these two aspects are dealt with separately.

4.1 B-Cell Recognition

4.1.1 The Hemagglutinin

It has been known for a long time that antibodies which neutralize the infectivity of influenza virus bind predominantly to the HA, and this is an important property which distinguishes the subtype.

Suggested locations of major antigenic sites on the HA molecule have been described using two complementary techniques: (1) the generation of a panel of monoclonal anti-HA antibodies for selecting viral mutants expressing antigenically changed HA molecules; the construction of an antigenic map of the HA molecule by analysis of the binding results; and (2) an analysis of the three-dimensional structure of HA molecules and a comparison of the amino acid sequences of the HA from related epidemic and mutant strains (WEBSTER and LAVER 1980; WILEY et al. 1981; GERHARDT et al. 1981). There is general agreement that these approaches show the presence of four antigenic determinants which have been variously designated as: site A (loop), a.a. 140–146; site B (tip), 187–196; site C (hinge), a bulge in the tertiary structure at the disulfide bond between Cys 52 and Cys 277; and site D (interface), which is in the interface regions between subunits in the HA trimer. The possibility that variation in one locus may be of greater advantage to the virus has been suggested (WILEY et al. 1981).

Recently, a panel of 16 monoclonal antibodies was tested for their binding to different physical forms of the HA molecule, prepared from A/Memphis/1/71 by bromelain digestion (NESTOROWICZ et al. 1985a). The antibodies (four per group) were classified according to their binding to sites A, B, C, or D of the HA molecule. The HA was in three forms – as the trimer, as the monomer or as a reduced and alkylated form. One or more of each group of four antibodies was able to neutralize the virus infectivity. It was found that (1). No antibodies bound to the denatured preparation. (2) Groups of antibodies directed to site A and site D clearly bound preferentially to the trimer compared with the monomer. (3) Antibodies which recognized site B varied in their binding properties – three bound almost equally well to the trimer and the monomer, and one bound only to the trimer. (4) Three antibodies which recognized site C clearly bound preferentially to the monomer. The fourth in this group bound almost equally well to the monomer and trimer. (5) Of the seven antibodies which had been tested for and found to have neutralizing activity, five bound preferentially to the trimer and the other two to the monomer. These results, which seem to agree with findings in some other systems, indicate that antibodies, including neutralizing antibodies to important antigenic sites of a viral surface antigen, react with conformational determinants, some of which may be to quaternary structures and therefore are nonlinear (discontinuous) amino acid sequences.

A number of groups have synthesized short peptides corresponding to sequences in the HA molecule, induced antibody formation to these peptides, and tested the ability of these antibodies to react with the HA molecule (e.g., JACKSON et al. 1982; MÜLLER et al. 1982; GREEN et al. 1982; SHAPIRA et al. 1984; NESTOROWICZ et al. 1985b). The peptides aimed to mimic regions of HA postulated to form important antigenic sites of the molecule or to represent most of the HA1 sequence. The results varied. NESTOROWICZ et al. prepared eight peptides varying in length from 9 to 25 amino acids long, but found only some of them to be immunogenic. Antibody raised to only one (the C-terminus of HA1, 24 amino acids in length) bound to intact virus. In contrast, GREEN et al. reported that most of their synthetic peptides, representing some

75% of the HA1 sequences, including sequences not present at the molecule's surface could elicit antibodies which bound to HA. Perhaps unexpectedly in view of these results, antibody raised to the intact HA did not bind to any of the peptides. The reason for these apparently contradicting results is not certain, but in testing the anti-peptide antibody for reaction with HA, the HA was attached to the plastic plate, where it may have assumed a conformation different from that in solution (e.g., ADA and SKEHEL 1985). Findings also differ on the ability of an N-terminal peptide of the HA2 molecule to induce the formation of antibody which binds to intact virus (ATASSI and WEBSTER 1983; NESTOROWICZ et al. 1985b). SHAPIRA et al. (1984) found that antibodies raised against a synthetic "loop" peptide (residues 139–146) did not react with intact virus but that antibodies to the longer peptide (residues 138–164 and 147–164) did (see also Sect. 10).

WYLIE et al. (1982) have shown that a proportion of monoclonal antibodies raised to infected murine spleen cells or infected cultured cell lines are specific for viral antigen, only if presented on the cell in association with the MHC gene product of the cells. This was not a rare event. Though some may have recognized nonstructural proteins, it appeared most recognized some viral antigen-cell complex, showing not only the capability of antibodies to recognize complex antigenic structures (discontinuous sequences?), but suggesting that these complexes might be highly immunogenic.

4.1.2 The Neuraminidase

Similar approaches to the above are being used to study the antigenic properties of NA. Variation in amino acid sequence has been observed to occur in regions which form a nearly continuous surface at the top of a subunit. The catalytic site, which is a large pocket of the distal surface, thus appears to be surrounded by variable regions. Studies with monoclonal antibodies to laboratory-derived mutants indicate that these areas with variable sequences are also antigenic (COLMAN et al. 1983).

4.1.3 Other Virion Proteins

The internal viral proteins of the virus have not been examined to the same extent as the two surface glycoproteins. Five monoclonal antibody preparations, each reacting to a different nucleoprotein (NP) determinant, were tested on a large number of viruses (infecting humans and animals) of different serological subtypes and isolated between 1930 and 1978 (Van WYKE et al. 1980). Several recombinant strains were also tested. One antibody detected an apparently invariant region while the other four detected variable regions. These authors concluded that point mutations and genetic reassortment contributed to the antigenic variability of NP but that these changes occurred independently of those in the surface glycoproteins.

Rather similar findings have been made for matrix protein (M) (Van WYKE et al. 1984). Monoclonal antibodies specific for M were reacted with five influenza A strains. Two epitopes showed antigenic variation, whereas the third appeared to be invariant. At least two of the three epitopes were in nonoverlapping domains.

The other proteins of the virion have not been examined to the same extent. Examination of NS1 products with polyclonal antisera showed some antigenic variation, but there was always substantial cross-reactivity (SHAW et al. 1982). The polymerase proteins show a very high (>96%) conservation of amino acid sequences so that substantial antigenic variation seems unlikely.

In summary, the advent of rapid sequencing techniques and monoclonal antibodies has allowed the accumulation of a very large amount of information about the HA molecule so that it is a well-characterized protein. The drive to obtain similar information for neuraminidase and especially for the other virion proteins has not been so great.

4.2 T-Cell Recognition

4.2.1 Regulatory T Cells

Despite the difference in antigen presentation to B and T cells, most notably in the context that antigen presentation to T cells requires some type of association with gene products of the MHC, it was unconsciously assumed by many that both T and B cells would recognize the same peptides. The past few years have seen a change in this attitude as information has been obtained that some peptides are more effective at stimulating T cells than reacting with B cells, and vice versa. Evidence for this now comes from several proteins so that it now should be regarded as a general finding. It has been shown by MANCA et al. (1984) and ALLEN et al. (1985) for hen's egg lysozyme, by STREICHER et al. (1984) for myoglobin, and, most recently, by MILICH et al. (1986) for the Pre-S-region of the hepatitis B surface antigen (Pre-S-HbsAg). These latter authors have shown that in mice the Pre-S-HbsAg is considerably more immunogenic than the HbsAg. The immune response to a synthetic peptide 120–145 from the Pre-S region is MHC restricted and contains two regions; segment 120–132 (as well as 120–145) stimulates T cells, whereas the segment 131–145 fails to do so, but the latter will bind to antibody against Pre-S-HbsAg. This not only illustrates the existence of different reactivities of peptides to B and T cells, but also shows that they may be adjacent in the protein molecule.

Numerous groups have studied the stimulation of Th cells by influenza virus hemagglutinin (e.g., ANDERS et al. 1981; THOMAS et al. 1982; REISS and BURAKOFF 1981) and, recently, it has been demonstrated that both subtype-specific and cross-reactive T cells can be stimulated by both HA1 and HA2 polypeptide chains, but that the latter is more important for stimulation of cross-reactive Th cells (KATZ et al. 1985). LAMB et al. (1982) examined the ability of human T-cell clones raised to the HA of an influenza A virus (A/Texas/1/77; H3N2) to respond in proliferation assays to peptides synthesized according to the amino acid sequences of a related strain X-47. All 12 peptides examined caused some proliferation, but peptide 20, a 25 amino acid segment at the carboxyterminus of HA1, was most immunogenic. This peptide is distant from the four proposed antibody-binding sites involved in viral neutralization. HACK-ETT et al. (1983) similarly identified a synthetic nonapeptide (residues 111–119) which stimulates a murine helper T-cell line from a responder mouse strain to the HA of A/PR8/8/34 (H1N1). It was, however, very much less efficient than the intact HA. It is located at the globular head region and again is distinct from the four proposed antibody-binding sites. HURWITZ et al. (1984) raised several HA-specific T-cell hybridomas and tested their ability to respond in proliferation assays to 43 antibody-selected PR8 virus mutants with known amino acid substitution differences. In this way, the hybridomas were placed into three specificity groups. One group responded to the nonapeptide 109–119, another group to the sequence 302–313 (corresponding to the finding of LAMB et al. 1982), and the third to a sequence including residue 136.

A second aspect, potentially of great importance, is the realization that T-cell antigenic sites tend to have amphipathic structures, that is, structures with separated hydrophobic and hydrophilic surfaces, with the hydrophobic residues displaying periodicity. ALLEN et al. (1985) showed this structure for a 10-mer peptide from hen's egg lysozyme (HEL) and STREICHER et al. (1984) for a CNBr fragment, 132–153, from sperm whale myoglobin (SWM6). DE LISI and BERZOFSKY (1985) have now studied the properties of 12 sites from 6 different proteins, including influenza HA1, which are antigenic for T cells; their method of analysis indicates that the amphipathic periodicity hypothesis is valid for 10 of these, including the two well-defined determinants of HA1. The converse – that if an amino acid segment is amphipathic it will be antigenic for T cells – does not necessarily follow. But the findings have potential for future vaccine strategy, as will be discussed in Sect. 10. It should be further noted that BABBIT et al. (1985) have now demonstrated specific binding of the HEL 10-mer peptide to the I-A glycoprotein from responder mice but not to the corresponding glycoprotein from nonresponder mice, some 12 years after MHC restriction between stimulator cells and responder T cells was first demon-strated (ROSENTHAL and SHEVACH 1973).

There have been several reports of T-cell clones which recognize viral pro-teins other than HA. Thus, LAMB et al. (1982) derived a panel of 11 human T-lymphocyte clones against A/Texas/1/77 (H3N2) and determined their speci-ficity as follows: NA, five clones; M, four clones; HA, one clone; NP, one clone. None responded to a B-strain influenza virus. The clones had an HLA restriction pattern which was consistent with helper activity. STERKERS et al. (1985) studied five human T-lymphocyte clones and one long-term culture for viral specificity. They had helper and cytotoxic activity and the phenotype of class II antigen restricted cells but the MHC specificity was not reported. Three were cross-reactive, one recognized NA, and two recognized HA. RUSSELL and LIEW (1979) reported experiments suggesting that helper T-cell activity to a viral internal antigen (M) could provide help for a subsequent HA antibody response. Mice primed to spikeless viral particles or to M and challenged later with intact virus gave an enhanced anti-HA response and the results were dis-cussed in terms of associative recognition in which "principal" and helper deter-minants were carried by different proteins. FISCHER et al. (1982) later showed that a human helper T-cell line raised to M could provide help in tissue culture for anti-HA antibody production by B cells in the presence of whole virus. THOMAS et al. (1982), using a different approach, were unable to show that M was a candidate for recognition by cross-reactive T-helper cells. Although RUSSELL and LIEW and FISCHER et al. tried to eliminate contaminating HA in their M preparations, this has been suggested as a reason for these findings. Further work needs to be done to establish the extent to which associative recognition may occur in this system.

Finally, there is very little information available about antigen recognition by suppressor T cells. LAMB et al. (1983) showed that brief incubation of cloned helper T cells with high concentrations of the appropriate synthetic peptide would render the cells unresponsive to a subsequent incubation with an "immunogenic" dose of antigen. The dose of peptide needed to achieve this effect was sufficiently high (30–300 ng/ml) to doubt whether such a mechanism could operate in vivo. LAMB and FELDMANN (1982) also used an ingenious approach, based on the idiotypic network theory, to generate a human suppressor T-cell clone to an autologous helper T-cell clone specific for M. The authors suggest that such mechanisms may operate in vivo and their further work in this area is awaited with interest.

4.2.2 *Effector T Cells*

Our knowledge about antigen recognition by effector T cells comes almost entirely from recent studies of one subset of T cells – cytotoxic T (Tc) cells. Such cells are generated in vitro or in vivo by exposure of a responder T-cell population to a stimulator cell expressing, it is believed, a viral antigen associated with an MHC antigen at the cell surface. The effector activity of the T cell is estimated in vitro by measuring the cytolytic activity on an infected, MHC class I-compatible target cell. The pattern of association required to render a target cell susceptible to lysis is generally considered to reflect a similar pattern on the presenting cell responsible for the induction of the T-cell response. Thus, the specificity of recognition can be studied at the induction or effector stage, but has been mainly done at the latter.

A variety of methods has been applied. The most direct involve the treatment of inducer or target cells with one of three techniques: (1) transfection with known DNA sequences coding for particular proteins; (2) infection with *recombinant* (hybrid) viruses which express the protein coded for by the inserted DNA; and (3) fusion with preparations, usually liposomes, containing defined proteins. Another approach has been to compare parental viruses with reassortant or mutant strains and this may implicate a specific viral protein. A different approach is to see whether antiserum to a particular protein interferes with effector activity. If a positive result is obtained with a polyclonal serum, stringent attempts should be made to show the specificity of the serum (e.g., REISS and SCHULMAN 1980a). The possible danger of a false-positive result using a polyclonal antiserum is overcome if a positive result is obtained with several monoclonal antibodies; however, if these give a negative result, it does not necessarily mean that a particular antigenic determinant is not present or a particular protein is not expressed, as evidence is now appearing that (1) different proteins, e.g., M1 and M2, may share amino acid sequences or that transfection with a short segment only of DNA coding for a viral protein may render a target cell susceptible to specific lysis (e.g., TOWNSEND et al. 1985) and (2) target cells transfected with DNA in this way and susceptible to lysis may not display the antigen at the cell surface (ibid.).

Shortly after the first demonstration that preparations of Tc cells could be generated against influenza virus (CAMBRIDGE et al. 1976; YAP and ADA 1977), several groups demonstrated that Tc cells raised to one A strain virus could lyze target cells infected with any A strain, but not a B-strain virus (DOHERTY et al. 1977; ZWEERINK et al. 1977; BRACIALE 1977). Limiting dilution analysis subsequently showed that the majority of Tc cells were "cross-reactive" (OWEN et al. 1982) and work with murine T-cell clones showed that they could be divided into three groups: those that were specific for the HA of the stimulating virus; those which recognized HA of all viruses of a given subtype; and those which recognize target cells infected with any A strain virus (BRACIALE et al. 1981).

In Table 1, results are assembled which indicate the viral proteins recognized by Tc cells. The field is moving rapidly, so the table will be incomplete at the time of publication.

Clearly, many investigators expected that Tc cells would predominantly recognize surface glycoproteins and particularly HA in view of the importance of this antigen in viral attachment to cells. There can be little doubt that in both murine and human studies HA is recognized by Tc cells, but the evidence now suggests (1) about 10%–15% of Tc precursor cells recognize HA and (2) there is some cross-recognition between H1 and H2, but very little, if any, between H2 and H3. Little is known about the epitope(s) recognized by Tc cells. WABUKE-BUNOTI and FAN (1983) reported that HA2 could induce a weak secondary Tc cell response, which was subtype specific, but neither BECHT et al. (1984) nor YAMADA et al. (1985) found HA2 prepared by different means had such activity. However, the latter authors found that a hybrid protein composed of the first 81 amino acids of NS1 and the complete HA2 subunit would stimulate a specific, Tc-cell response (see below). To date, there are no reports unequivocally demonstrating that NA is recognized by Tc cells, though this is likely to be the case (STERKERS et al. 1985).

In view of such findings, experiments were carried out to establish whether other viral proteins were recognized by Tc cells. KEES and KRAMMER (1984) using limiting dilution analysis of early responses to reassortant viruses found that up to 90% of Tc cells in C57Bl/6 mice recognized internal proteins. Are all proteins recognized? At the time of writing, the *published* literature indicates HA, NA, NP, M, and PB2 (see Table 1). Two important factors are emerging, however: (1) Variation between humans and between mouse strains in the recognition of different antigens. This is discussed later in this section. (2) Variation between individual animals within an inbred mouse strain in the recognition of a given antigen.

YEWDELL et al. (1985) claim a "significant" proportion of cross-reactive Tc cells in BALB/c mice recognize NP. ANDREW et al. (1986) using limiting

Table 1. Viral proteins recognized by influenza virus-specific cytotoxic T cells

Proteins	Cellular expression		Spe-cies[a]	Me-thod[b]	References
	Stimu-lator	Target			
Structural					
Hemagglutinin (HA)	+	+	H, M	T, RV, F, Ab	WRAITH and ASKONAS (1985), FLEISCHER et al. (1985), STITZ et al. (1985), BENNINK et al (1984), KOSINOWSKI et al. (1980), BRACIALE et al. (1984), BRACIALE (1979), TOWNSEND et al. (1984a), EFFROS et al. (1979), ASKONAS and WEBSTER (1980), YAMADA et al. (1985), ANDREW et al. (1986)
Neuraminidase (NA)		+	H		STERKERS et al. (1985)
Matrix Protein (M1)	+ −	+	H, M	Ab, RM RV	REISS and SCHULMAN (1980), FLEISCHER et al. (1985), J. BENNINK (personal communication)
Nucleoprotein (NP)		+	H, M	T, RV RM	McMICHAEL et al. (1986), YEWDELL et al. (1985), FLEISCHER et al. (1985), TOWNSEND et al. (1984a, b), TOWNSEND and SKEHEL (1984)
Polymerases					
PA		+	M	RV	J. BENNINK (personal communication)
PB1		+	M	RV	J. BENNINK (personal communication)
PB2		+	M	RM, RV	BENNINK et al. (1982)
Nonstructural					
Matrix protein (M2)		−	M	RV	J. BENNINK (personal communication; preliminary data)
NS1		+	M	RV	J. BENNINK (personal communication); see also YAMADA et al. (1985)
NS2		−	M	RV	J. BENNINK (personal communication; preliminary data)

[a] H, human; M, mouse
[b] T, transfection; RV, recombinant virus; RM, reassortant or mutant virus; F, cell fusion; Ab, inhibition by antibody

dilution to determine precursor frequency find about 30% recognizing NP. PALA et al. (1986), using L cells transfected with NP and D^b genes, estimated the proportion of A virus cross-reactive Tc specific for NP. They found a great variation in the frequency of NP-specific Tc between individuals in an inbred mouse strain, with some showing no recognition of NP, though having a strong A virus-specific cross-reactive Tc response. It has also been pointed out that there are at least two classes of NP-specific Tc, one which is fully cross-reactive between all A strains and another which distinguishes NP in two groups of type A viruses, isolated between 1934–1943 and 1943–1979 (TOWNSEND and SKEHEL 1984). There may be three determinants on the NP molecule recognized by Tc (TOWNSEND et al. 1985). Bennink and colleagues (J. Bennink, personal communication) have constructed vaccinia virus recombinants containing the

DNA coding for other viral antigens and have found that all three polymerases are recognized by murine Tc cells. A reasonably safe prediction (e.g., STITZ et al. 1985) would be that if sufficient mouse strains and a large enough number of humans are examined, it will be found that most if not all viral proteins may be recognized.

Are nonstructural viral antigens – M2, NS1, and NS2 – recognized by Tc cells because it is clear that M2 and NS1 are well expressed at the infected cell surface, a finding which per se would suggest that these antigens are candidates for recognition? Bennink and colleagues (J. Bennink, personal communication), using recombinant vaccinia preparations, find that NS1 but not M2 is recognized by Tc cells. In contrast, YAMADA et al. (1985) found that an *E. coli*-produced NS1 protein did not induce Tc cell generation, whereas a hybrid HA2-part NS1 molecule did. This result, together with that of WABUKE-BUNOTI and FAN (1983), does suggest the presence of a subtype-specific epitope for Tc cells on HA2.

In contrast to the extensive work with effector cells expressing Tc-cell activity, little work has been done on effector cells expressing DTH, except to show that populations which are specific to one subtype or are cross-reactive may be produced (e.g., ADA et al. 1981). Preparations may be class I or II MHC antigen-restricted. In view of the fact that cloned T cells may show both cytotoxic and DTH activities (e.g., LIN and ASKONAS 1981) and cells expressing cytotoxic activity may be class I or II MHC antigen-restricted (LUKACHER et al. 1985), the antigenic recognition patterns seen with Tc cells probably apply also to cells expressing only (detectable) DTH activity.

It should also be noted that the activation of precursor effector T cells may be affected by *MHC-restricting elements*. Thus, using influenza-specific cloned Tc cells and infected, parental and mutant H-2^b cells, one K^b-restricted clone was found not to recognize an infected K^b mutant bM1 target cell and all 15 D^b-restricted clones did not recognize infected, D^b mutant bM14 target cells.

These effects are becoming even more pronounced when recognition of individual viral proteins is considered. Bennink and colleagues (J. Bennink, personal communication) using vaccinia recombinants have found that all three viral polymerases are recognized by Tc cells, but that the recognition is influenced by the H2 haplotype of the mouse strain used, particularly in the case of H-2^b, H-2^k, and H-2^b strains. PALA et al. (1985) using transfected target cells have found that in C3H-H-2^{02} (K^{d}D^k) mice, D^k is a low-responder allele for NP recognition by Tc cells. This observation was extended using vaccinia recombinants to show that D^{d}L^d and K^b were also low-responder alleles in conjunction with NP (B.A. Askonas, personal communication). It was found previously for a number of viruses that Tc-cell responses can show predominant restriction to either K or D region molecules, e.g., DOHERTY et al. (1978) reported that H-2^b mice primed intraperitoneally to influenza were unable to mount a K^b-restricted response in the presence of a D^b-restricted response and a variety of explanations has been proposed. PALA and ASKONAS (1985) have found that in H-2^b mice the magnitude of the K^b-restricted anti-influenza responses is affected both by the site of infection and the type of stimulator cell. Thus, a variety of factors can contribute to low responsiveness. Despite this, an overall assessment is that MHC class I genes function as Ir genes. Though there are few firm data available, this would be expected to occur in outbred populations, such as humans. It may be hypothesized that in humans poor or nonresponders to influenza virus infections may have been eliminated in earlier outbreaks or, perhaps more likely, MHC polymorphism and the (now recognized) large number of viral antigens recognized by both Th and Tc cells would minimize such possibilities (for review of this topic, see TOWNSEND and MCMICHAEL 1985). It may be possible to test for such Ir gene effects in a population where the incidence of memory Tc cells has over the years declined markedly, e.g., in Oxford (MCMICHAEL et al. 1983) on the next occasion there is a major outbreak of influenza.

4.3 Antigenic Drift and Shift

The two surface antigens of influenza A viruses undergo two types of antigenic change. Antigenic drift is the term describing minor changes in the HA and NA; antigenic shift involves major changes in these two surface glycoproteins. This topic has been the subject of intense investigation and was reviewed recently (MURPHY and WEBSTER 1985).

Antigenic drift in both the HA and NA occurs by point mutation in the genes, resulting in an accumulation of amino acid sequence changes. In the case of the HA of A/Hong Kong/1/68 and later H3N2 viruses isolated between 1968 and 1977, most changes occur in the HA1 molecule, with three alterations occurring in HA2 (BOTH et al. 1983). The changes in HA1 clustered around the four main antigenic sites have been described earlier in this section. Drift can be mimicked in the laboratory by growing virus in the presence of monoclonal antibodies. Variants of both HA and NA occur at a frequency of about 10^{-5} and do not bind to the selecting antibody. When tested with polyclonal antisera, antigenic variants with only one amino acid change seem to have little effect on the total antigenic properties of the HA or NA, suggesting that a significant change from an epidemiological point of view requires changes in two or more sites. A plausible explanation is that mutations occur sequentially during the spread of the virus, suggesting that a change in only one epitope can give some survival advantage.

Three mechanisms have been proposed to explain antigenic shift. The change which occurred in 1968 when H3 replaced H2 is so large that it has been proposed that the H3 human virus could only have been derived by genetic reassortment between human and animal or human viruses. Viral genetic reassortment between humans and between humans and lower animals has been observed and several studies involving genetic and biochemical analysis support this mechanism for the appearance of H2 and H3 strains. In contrast, the strain of H1N1 that appeared in northern China in 1977 and spread worldwide is so similar in all genes to the virus which caused an epidemic in 1950 that it is reasonable to propose that the earlier virus had remained dormant for the 27-year interval, e.g., in a frozen state. The third possibility is that an animal or bird virus changed and became infectious for humans. An example is the isolation of identical influenza viruses from pigs and humans on the Wisconsin farm in the mid-1970s (HINSHAW et al. 1978).

The critical aspect is that there is no accepted method available for predicting the precise changes which occur in either of these processes. The potential hazard this represents to animal and bird life is demonstrated by the high mortality which occurred in seals in 1980 (GERACI et al. 1982) due to a mixed infection with avian viruses. The resulting virus, A/seal/Mass/1/80 (H7N7), was infectious but of low virulence for humans, but this episode does indicate the potential of avian influenza viruses to infect and cause disease in mammals.

5 Patterns of Infection

5.1 Influenza in Humans

Human influenza is primarily an infection of the upper respiratory tract and major central airways. The pathology, characterized by desquamation of the epithelium, involves the nasal mucosa, larynx, and tracheobronchial tree. Viral antigen may be detected by immunofluorescence in the epithelial cells and mononuclear cells and may persist for several days after virus is no longer recoverable (SWEET and SMITH 1980). Influenza pneumonia occurs rarely and is characterized histologically by a desquamative interstitial pneumonitis. Viral antigen may be detected, in the cases of pneumonia, in both type I and type 2 pneumocytes and alveolar macrophages. Although physiological abnormalities of small airways function may be demonstrated in uncomplicated influenza (HALL and DOUGLAS 1980), these abnormalities relate to airway hyperreactivity and do not necessarily reflect local infection.

The major method of transmission of influenza virus is airborne; direct spread via infectious droplets is less common (MURPHY and WEBSTER 1985). Inhalation of infected aerosols may result in the deposition of virus throughout the respiratory tract, including the alveoli. It is unclear why there is subsequently less pathological involvement at the alveolar level. Influenza virus may be more efficiently cleared by local defense mechanisms, especially the alveolar macrophage, preventing alveolar infection. Alternatively, the alveolar epithelium may be less susceptible to infection as suggested by in vitro studies (ROSZTOCZY et al. 1975).

5.2 Influenza in Animals and Birds

Influenza infection occurs naturally in many avian species and in some mammals including pigs, horses, seals, and mink. The study of infection in these hosts has provided valuable data on the epidemiological links between avian, mammalian, and human influenza viruses, particularly on the basis of antigenic shift of human influenza viruses. However, studies on the immune response have been largely restricted to experimental infections in mice and ferrets.

The mouse has been the most widely used animal for experimental studies of influenza virus infection. Its advantages over other animal models, aside from the similarities between murine and human influenza pneumonia, include the ease of breeding and availability of inbred strains, the extensive background knowledge of the murine immune system and the MHC, and the availability of well-defined reagents for characterizing the components of the immune response. The pattern of infection in mice depends on the degree of host adaptation of the virus. Infection with unadapted strains generally does not produce overt disease, unless very high inocula are used, although virus may replicate in the lungs, bronchi, and nasal mucosa. Infection with mouse-adapted strains, either intranasally or by aerosol, results in severe disease with alveolar involvement as the predominant finding. The pulmonary pathology in mice is similar to that seen in influenza pneumonia in humans. Lung viral titers peak within 3–4 days and decline to undetectable levels by 10 days, although viral antigen may be detected, by ELISA, up to 2 months after infection (ASTRY et al. 1984).

Intranasal inoculation of ferrets with influenza virus results in a similar disease to uncomplicated human influenza, with a similar distribution of viral antigen in the respiratory tract. There is little evidence for alveolar involvement even during infection with virulent clones (SWEET et al. 1981). These observations are consistent with the demonstration that ciliated epithelial cells from ferrets are better able to support virus replication and to release virus than are alveolar cells (CAVANAGH et al. 1979). The ferret respiratory tract, as opposed to that of the mouse, is more feasibly dissectable into upper and lower divisions, allowing a more selective approach in studying the immune response, particularly the antibody response, in the respiratory tract. However, the ferret is limited as a model for studying the immune response, primarily by lacking the advantages of the murine model as well as by the difficulty and expense in maintaining sufficient numbers of ferrets for detailed experiments.

6 Nonspecific Responses

The nonspecific immune system in influenza infection has been evaluated either by delineating the components of the system separately – namely fever, macrophages, natural killer (NK) cells, interferon, and complement – or by examining the system collectively and independently of the specific arm of the immune response. The complex interactions within this system and with the specific arm of the immune response, however, make it difficult to dissect the roles of the respective components in recovery from influenza infection. The response of these components in influenza infection, and where possible their role, will be reviewed. It is not feasible to consider the details of the interacting functions of these components in antiviral immunity in this review. Their interaction with the specific components of the immune response in influenza will be considered later.

6.1 In Experimental Influenza

In experimental influenza the nonspecific immune response is implicated in limiting viral spread and initiating recovery. Several observations have been made which tend to substantiate this role. Firstly, pretreatment of mice with *P. acnes* (*C. parvum*) before a lethal virus challenge results in lower lung viral titers and lower mortality (MAK et al. 1983; GANGEMI et al. 1983). This protective effect correlates with increased lung interferon levels, macrophage content, and NK cell activity in the absence of any demonstrable change in either T- or B-cell function. Secondly the survival of athymic mice following a sublethal infection with influenza virus is dependent on either the nonspecific immune response or IgM antibodies (YAP and ADA 1979). The role of antibody in recovery from infection is considered later. The persistence of virus in the lungs of surviving mice for prolonged periods suggests, however, that neither mechanism is sufficient for complete recovery. Finally, the early decline, during the 2nd and 3rd day, of nasal virus titers in infected ferrets precedes the development of a measurable specific immune response and correlates with the degree of preceding pyrexia and inflammatory cellular infiltrate (TOMS et al. 1977). It is difficult to assign responsibility to any one component of the nonspecific immune system in these observations. Descriptive data on the components provide further insight into the mechanisms responsible for initial recovery.

The role of fever as a host defense mechanism is suggested by the correlation between the height of fever and subsequent rate of decline of nasal virus titers in ferrets. Furthermore viral replication in nasal turbinate organ cultures is inhibited at pyrexial temperatures (SWEET et al. 1978) and suppression of fever in ferrets by nonpharmacological means results in delayed clearance of virus (HUSSEINI et al. 1982).

Alveolar macrophages from normal murine lungs are susceptible to infection by influenza virus in vitro (RODGERS and MIMS 1981; WELLS et al. 1978). Although the virus undergoes abortive replication only, infectious virus, possibly on the surface of macrophages, may form infectious foci when cocultured with susceptible cells. Alveolar macrophages recovered from influenza-infected mice acquire resistance to in vitro infection which is induced by interferon (RODGERS and MIMS 1982b). Macrophages recovered from influenza-infected lungs may also mediate lysis of infected cells independent of antibody (MAK et al. 1982a). Influenza-infected macrophages also produce interferon and may act as antigen-presenting cells (WYDE et al. 1982; LYONS and LIPSCOMB 1983). The effect of influenza virus infection on release of interleukin-1 has not been studied.

Natural killer cells may play a role in limiting viral spread by lysis of virus-infected cells and production of interferon (WELSH 1981). Enhanced levels of natural killer cells can be detected in pulmonary lymphocytes 48 h after influenza infection in mice (LEUNG and ADA 1981a; STEIN-STREILEIN et al. 1983).

Pulmonary levels of type I (alpha-, beta-) interferon rise rapidly during murine influenza infection and correlate directly with the degree of viral replication (WYDE et al. 1982). Alveolar macrophages and lymphocytes recovered early from infected lungs are the major sources of interferon released in vitro. In infected ferrets, nasal levels of interferon also rise early and correlate directly with nasal virus titers (HUSSEINI et al. 1981). Intranasal instillation of anti-interferon serum results in an increase in viral titers and an increase in host mortality (HOSHINO et al. 1983), whereas intravenous administration has no affect on the course of influenza infection (GRESSER et al. 1976). In mice bearing the Mx gene an increased resistance to influenza virus, attributable to enhanced sensitivity to alpha/beta interferon, can be abolished by pretreatment with anti-interferon serum (HALLER 1981). These observations indicate a definite role for interferon in limiting viral spread. The role of gamma interferon is considered later (Sect. 8).

Several in vitro observations suggest possible mechanisms by which complement may be activated during influenza infection in the absence of antibody. Virus particles themselves, virus-infected cells, and virus-induced desialation of cells may activate complement by either the classical or alternate pathways (LAMBRE et al. 1983). Subsequently binding of complement in the absence of specific antibody may neutralize virus or result in lysis of virus particles. The importance of complement in influenza infection is shown by the increased mortality in mice depleted of complement and in C5-deficient mice, although the major protective effect of complement was mediated late in infection and probably through antibody-dependent lysis of infected cells (HICKS et al. 1978).

6.2 In Human Influenza

While there are no data in humans comparable to that in experimental influenza which support a role for the nonspecific immune system in recovery from influenza, descriptions of the responses in humans are consistent with those in animal models.

Alveolar macrophages recovered from human bronchoalveolar cell preparations are susceptible to influenza in vitro, though replication is abortive (RODGERS and MIMS 1982a). Alveolar macrophages infected in vitro produce alpha-interferon and retain the ability to act as accessory cells (ETTENSOHN and ROBERTS 1984).

Large granular lymphocytes (LGLs) separated from peripheral blood lymphocytes may be stimulated in vitro by influenza virus (DJEU et al. 1982), resulting in enhanced natural killer activity and production of alpha- and gamma-interferon. Furthermore, LGLs provide an accessory function in the generation of influenza-specific cytotoxic T cells through unidentified soluble factor(s) (BURLINGTON et al. 1984). Enhanced natural killer activity may also be detected in circulating lymphocytes early after influenza infection in volunteers (ENNIS et al. 1981).

Alpha-interferon is detected in acute-phase sera during natural infection (ENNIS et al. 1981; Green et al. 1982). Local nasopharyngeal interferon levels peak early after artificial challenge and correlate directly with nasal virus titers (MURPHY et al. 1973). In vitro stimulation of peripheral blood lymphocytes results in the production of both alpha- and gamma-interferon, higher levels of gamma-interferon being produced in recently vaccinated volunteers (ENNIS and MEAGER 1981).

In summary, the nonspecific components of the immune response to influenza virus infection appear to have a role in limiting viral spread and initiating recovery prior to the development of T- and B-cell responses in primary infections. Although this proposition is based largely on data from animal studies, there is little reason to doubt that a similar role in human influenza does not occur. The validity of this proposition in human influenza may eventually be confirmed by observations on the course of influenza in persons with selective hereditary defects in the nonspecific components of the immune response.

7 The Antibody Response

The antibody response to influenza infection in humans has been thoroughly described although some observations from animal studies yet to be demonstrated in man will also be considered. In contrast much of the data to be reviewed pertaining to the role of antibody in protection and recovery have been derived from animal studies.

7.1 Dynamics of the Response

7.1.1 The Anti-Hemagglutinin Response

The class-specific antibody response to the hemagglutinin subunit has been studied in persons challenged with live attenuated vaccines, using the ELISA technique. A characteristic primary serological response was observed in children experiencing a primary infection (MURPHY et al. 1982). Serum IgM and IgG responses occurred in all cases whereas IgA responses occurred less frequently and were less marked. IgA formed the major response in nasal secretions and occurred in most cases. IgG responses in nasal secretions occurred infrequently and to a lesser degree than both IgA and IgM. A secondary antibody response was detected in young adults primed by natural infection and challenged subsequently with a live attenuated subtype variant (BURLINGTON et al. 1983). Serum IgG and IgA responses occurred in most cases; there was a correlation between serum and secretory IgA responses but not IgG responses. IgM responses were detected infrequently in serum and not in nasal secretions. IgM responses during reinfection occur variably, presumably relating to the degree of antigenic variation between succeeding viruses (GONCHOROFF et al. 1982).

Serum HI titers gradually decrease over the first 6 months after infection and may then persist for several years as subsequent infections by related virus strains boost titers ("original antigenic sin"). The finding of antibody to H3 in 1968 in persons born before 1892, and to H1 in 1976 in persons born before the early 1950s, illustrates this (COUCH and KASEL 1983). In contrast, only about 30% of adults will have detectable neutralizing antibody in nasal secretions 1 year following intranasal immunization with an inactivated vaccine (WALDMAN et al. 1973). Nasal wash IgA was detectable, by ELISA, in children 3–6 months after natural infection but not 10–18 months after immunization with a live attenuated virus vaccine (WRIGHT et al. 1983).

Data on the antibody response in the lower respiratory tract of humans are limited. WALDMAN has demonstrated a threefold increase in HI antibody in bronchoalveolar lavage fluid following aerosol administration of an inactivated vaccine (WALDMAN et al. 1973). Following intranasal challenge with a low dose of live attenuated vaccine, an antibody response in the lower respiratory tract was observed in primed individuals and was predominantly IgG in nature (ZAHRADNIK et al. 1983). Antibody responses in all classes develop in mice within the 1st week, though titers in bronchial washings were markedly lower than those in serum (ZEE et al. 1979).

Antibody-producing cells (APCs) have been detected by measuring antibody production from cultured human peripheral blood lymphocytes. Immunization with either live or inactivated vaccines has been shown to result in the appearance of circulating APCs (YARCHOAN et al. 1981; MITCHELL et al. 1982). In experimental infection in ferrets and mice, APCs detected by the hemolytic plaque technique have been found in mediastinal lymph nodes and spleen (McLAREN and BUTCHKO 1978; REISS and SCHULMAN 1980b). Until recently, data on the response as measured by APCs in the lung have been negligible. Using an ELISA-plaque assay, we have detected influenza-specific APCs, producing IgM, IgG, and IgA, in murine lungs during primary infection, displaying similar kinetics to a primary serological response (P.D. Jones and G.L. Ada, manuscript in preparation).

7.1.2 Responses to Other Viral Proteins

Using sensitive assays, antibody to the neuraminidase subunit can be detected in serum during both primary and secondary infections and may subsequently persist for many years. The dynamics of anti-NA antibody formation and duration in nasal secretions in primed individuals after natural infection are similar to those for the anti-HA response, with IgA the major isotype (HRUSKOVA et al. 1976). Antibody responses to the matrix protein and nucleoprotein also develop frequently after infection. There are no reports of the antibody response in man to minor viral proteins though polyclonal antisera and monoclonal antibodies to the polymerase and NS1 protein have been prepared in animals (see Sect. 4.1.3).

7.2 Specificity of the Response

The specificity of antibody to the hemagglutinin subunit in the context of the broad antibody response following infection has been characterized using virus-adsorbed sera. The antibody population consists of antibodies to strain-specific determinants of the hemagglutinin of the infecting strain and of previous infecting subtype variants and cross-reacting antibodies to shared determinants of different subtype variants. The relative composition of the antibody response is largely dependent on the host's prior antigenic experience. After natural infection, the majority of unprimed children produce predominantly strain-specific antibodies to the infecting virus; cross-reacting antibodies constitute only a small component of the response (OXFORD et al. 1981). In contrast, the majority of adults, previously exposed to an earlier subtype variant, produce predominantly cross-reacting antibodies as well as strain-specific antibodies to the previously encountered strain (OXFORD et al. 1979).

These observations have been subsequently confirmed at the clonal level by analysis of the specificity of antibodies produced by stimulation of individual B-cell precursors in limiting dilution cultures (YARCHOAN and NELSON 1984). It was further shown that viruses of one subtype may stimulate the production of antibodies specifically directed to the HA of a different subtype or cross-reactive antibodies directed to shared determinants on either the HA or other proteins of different subtype viruses. The availability of monoclonal antibodies to known epitopes on the HA molecule now allows an examination of the response of individuals to particular epitopes, by testing the ability of their sera to compete with the monoclonal preparations for binding to the HA.

The specificity of the secretory antibody response is less well defined although similar patterns of a predominantly homotypic response with reports of a broadened heterotypic response dependent on previous antigenic exposure have been described (WALDMAN et al. 1970b; SHVARTSMAN et al. 1977). In some instances the specificity of the secretory response has been broader than that of the corresponding serum response.

7.3 Antiviral Effects of Antibody

7.3.1 Neutralization of Infectivity

It is generally considered that neutralizing antibody prevents entry of virus into susceptible cells by sterically blocking the association of virus with cellular receptors, as has been shown with reovirus (LEE et al. 1981). There has also been much argument over the kinetics of antibody-mediated viral neutralization (e.g., DELLA-PORTA and WESTAWAY 1978). Such information is important in considering how antibody affects viral neutralization and DIMMOCK and colleagues (POSSEE et al. 1982; TAYLOR and DIMMOCK 1985a, b) have recently studied this problem with respect to influenza virus, using different forms of antibody – IgG, IgM, and monomeric and secretory IgA. Profound differences were found between the monomeric (IgG, IgA), and oligomeric (IgM, sIgA) forms. The surprising initial finding was that type A influenza viruses, neutralized by polyclonal or monoclonal anti-HA IgG, attached to a variety of susceptible cell types, at temperatures between 4° and 37° C, with kinetics indistinguishable from those of nonneutralized virus. Furthermore, the kinetics of internalization of neutralized virus, its subsequent uncoating, and the transport of virion RNA to the cell nuclei was unchanged compared with infectious virus. Loss of infectivity seemed to result from the inhibition of a later stage in the replication cycle, thought to be inactivation of the viral transcriptase. Virus neutralized by monomeric IgA did not prevent attachment of the virus to susceptible cells, but its action was not investigated further.

In contrast, neutralizing IgM and sIgA prevented the attachment of up to half the virus to the cells and the portion which did attach was not internalized. These authors point out that virus-IgM complexes on the cell surface could well render that cell susceptible to C′-mediated lysis, but do not discuss the rationale of the different results obtained with the different antisera. Perhaps the binding of oligomeric forms of antibody, IgM, and sIgA simply interfered sterically with the availability of many HA "anti-receptors" to bind to the cell-receptor; the portion of the virus-IgM complex which binds does so rather poorly and certainly less well than virus neutralized by IgG or monomeric IgA. Two comments can be made: (1) The different modes of action of the various Ig isotypes may help to explain why this has been a controversial subject and (2) it would seem that nature has gone to considerable trouble to ensure that if antibody of the correct specificity is present, it will neutralize viral infectivity independent of the class of antibody involved.

7.3.2 Lysis of Virus-Infected Cells

Antibody may modify infection by restricting viral spread by lysis of virus-infected cells. This effect is mediated either by the lytic action of complement or by antibody-dependent cell-mediated cytotoxicity (ADCC).

Complement-mediated lysis of virus-infected cells proceeds through the alternate pathway of complement (PERRIN et al. 1976). Although the alternate pathway may be activated by virus-infected cells independent of antibody, lysis only occurs in the presence of specific antibody. Antibody mediating this effect is subtype specific and develops after influenza infection and vaccination (VERBONITZ et al. 1978).

Similarly, antibody responsible for ADCC appears after natural infection and vaccination (HASHIMOTO et al. 1983). Effector cells in human blood mediating ADCC include natural killer cells, neutrophils, and monocytes. The NK cells are the most efficient effector cells and require less antibody to produce cytolysis than the complement-mediated mechanism. As the responsible antibody primarily recognizes the HA molecule on the infected cell surface, ADCC is also subtype specific.

7.4 Protective Role of Antibody

7.4.1 Anti-HA Antibody

Evidence for the protective role of antibody is either based on demonstrating a correlation between antibody level and resistance to infection or, more directly, prevention or modification of infection mediated by passive transfer of antibody. Field and volunteer studies have shown that resistance to infection is correlated with serum anti-HA antibody levels. However, direct evidence that antibody is the mediator of protection has largely come from animal studies. Passive transfer of immune serum will protect recipients against homotypic challenge (VIRELIZIER 1975). This is more clearly seen in immunosuppressed recipient mice unable to mount an active host response. That antibody is essential for protection is demonstrated by the finding that mice selectively suppressed for antibody production can recover from infection, yet are susceptible to reinfection in the absence of RIA-detectable antibody (KRIS et al. 1985). A beneficial effect of seroprophylaxis in human influenza has also been claimed in studies cited by SHVARTSMAN and ZYKOV (1976).

Infection also confers a significant but lesser degree of immunity to subtype variants, although the extent to which this is due to antibody is unclear. The observations that heterotypic immune serum passively transferred in mice is less protective than homotypic serum (VIRELIZIER 1975), and that cross-reacting antibody to shared determinants on subtype variants is less efficient in neutralizing virus in vitro (HAAHEIM and SCHILD 1980), are not inconsistent with a role of antibody in modifying infection.

Epidemiological observations in humans also indicate that infection with one subtype does not confer immunity to other subtypes. In contrast heterotypic immunity between subtypes modifying infection may be demonstrated in experimental influenza and is associated with an enhanced antibody response to the challenge virus. However, mice are not protected by the passive transfer of immune serum prior to challenge with a different subtype virus (VIRELIZIER 1975).

7.4.2 Relative Importance of Serum and Secretory Antibody

To be protective, anti-HA antibody must be present at the mucosal surface, having been either produced locally or derived from serum. Resistance to infection in humans has been correlated with anti-HA in nasal washings of either the IgG or IgA isotypes (COUCH et al. 1981; CLEMENTS et al. 1983). Further evidence that IgG is protective has been inferred from the inverse relationship between transplacentally acquired antibody in infants and the frequency of influenza infection (PUCK et al. 1980). Maternally derived antibody in animal studies has also been shown to be protective though protection is mainly acquired during suckling (REUMAN et al. 1983; HUSSEINI et al. 1984). However, the protective effect of passively acquired antibody from immune mothers or in transfer experiments is limited to the lower respiratory tract, primarily the lung, in ferrets and mice – infection of the upper respiratory tract and trachea is not prevented. The protection afforded by serum-derived IgG to the lower respiratory tract is consistent with the greater proportion of IgG relative to IgA in lower respiratory tract secretions compared with the lower proportion of IgG in nasal washings (MCDERMOTT et al. 1982).

Immunoglobulin isotypes are distributed in the human respiratory tract in a similar pattern with the relative proportion of IgG to IgA increasing in the lower part. The degree to which serum-derived IgG contributes to prevention of infection may accordingly vary, depending on the site of virus deposition. IgA in nasal washings may assume an apparent greater importance in protection against virus challenge in human volunteer studies when virus is usually administered by intranasal droplets. When virus is transmitted naturally by aerosol, it is deposited throughout the respiratory tract and locally produced antibody,

IgA, and possibly IgG, with an additional effect from serum-derived IgG, primarily to the lower respiratory tract, prevent infection.

7.4.3 Role of Antibody in Recovery

Since patients with agammaglobulinemia recover from influenza infection, it is evident that antibody is not essential for recovery. However, the beneficial effect of hyperimmune serum in modifying disease in humans with influenza suggests an additive role for antibody in recovery (SHVARTSMAN and ZYKOV 1976), and this is supported by animal studies. Passive transfer of immune serum to normal mice early after infection had an additive effect in reducing lung virus titers (YAP and ADA 1979). However, passive transfer of antibody to infected nude mice, though it initially lowers lung virus titers and prevents dissemination, has a transient effect as virus shedding reappears in 1–2 weeks (ASKONAS et al. 1982a). By selectively suppressing the antibody response with anti-IgM, KRIS et al. have shown that mice can recover in the absence of antibody (KRIS et al. 1985).

7.4.4 Antibody to Other Viral Proteins

Antibody to the neuraminidase can neutralize viral infectivity in vitro or in vivo only in extremely high concentrations, probably due to steric hindrance of hemagglutinin-binding sites. However, smaller amounts of anti-neuraminidase antibody inhibit the release of virus from infected cells, possibly by cross-linking budding virus particles. The protective effect of anti-neuraminidase antibody in human influenza has been evaluated in field studies following the introduction of the H3N2 subtype and in challenge studies, with volunteers primed naturally or artificially by neuraminidase-specific vaccines, allowing an assessment of protection independent of an anti-HA antibody response. The presence of anti-NA antibody has been inversely correlated with infection rates and more consistently with a reduction in clinical illness (SCHULMAN 1975). Similarly, studies in animals have shown a protective effect of actively acquired anti-NA antibody, primarily in modifying infection rather than prevention. This effect can be reproduced by passive transfer of anti-NA serum though the effect is less pronounced than that of anti-HA antibody (VIRELIZIER 1975).

There is no evidence that antibodies to the matrix protein or nucleoprotein have a significant role in immunity to influenza. Mice passively immunized against these components are not protected and antisera to either component has no neutralizing activity in vitro (VIRELIZIER et al. 1976).

8 The Cell-Mediated Immune (CMI) Response

8.1 Cell Characteristics

It has already been mentioned in connection with antigen recognition (Sect. 4) that T-lymphocytes have in the past been classified according to their activities

Table 2. Characteristics of human and murine influenza-immune T cells

MHC antigen requirement for activation	Lymphocyte antigen expressed		Biological activities	Cell preparation (References)
	Human	Mouse		
Class I	T8		Cytotoxic Suppressor? DTH?	Secondary[a]
		Lyt2[+]	Cytotoxic	Primary[b], secondary[b], clones[c]
		L3T4[−]	DTH	Primary[b], secondary[b], clones[c]
		Lyt1[−]	Helper-inducer	Primary[b]
Class II	T4		Helper-inducer	Secondary[e], clones[f]
			Cytotoxic	Clones[g]
			DTH?	
		L3T4[+]	Helper-inducer	Primary, secondary[b]
		Lyt2[−]	DTH	Primary, secondary[b]
		Lyt1[+]	Cytotoxic	Secondary[h], clones[k]

The query indicates activities for those cell preparations which have not been measured in the influenza system

Representative references:
[a] McMichael et al. (1977); [b] e.g., Ada et al. (1981); [c] Lin and Askonas (1981); [d] Mizuochi et al. (1985); [e] Callard (1979); [f] Lamb et al. (1982); [g] Kaplan et al. (1984); Sterkers et al. (1985); [h] Morrison, quoted in [k]; [k] Lukacher et al. (1985)

as follows: regulatory T cells, Th and Ts, with helper-inducer and suppressor activity respectively; and effector T cells, Tc and Td, mediating cytotoxic and DTH activities. Table 2 lists the characteristics of T cells from human and murine sources with respect to their MHC requirements for activation, some surface markers commonly used to distinguish them, and the biological properties which cell preparations from primary responses (in vivo responses in naive hosts, e.g., mice) or secondary responses (usually in tissue culture with responder cells derived from primed hosts) or which cloned cells display. It is apparent that T cells occur in two major cell lineages according to (1) the MHC activation profile and (2) the presence of specific surface markers, Lyt2 and T3T4 in the mouse (Lyt1 differs only quantitatively) and T4 and T8 in humans. In contrast, work on influenza virus-specific T cells has been largely responsible for showing that a particular cloned cell line may mediate a number of activities, e.g., in the case of class I MHC restricted cells, Tc and Td activities. Though it has not been formally shown yet, it may well be found that one clone of class I or II MHC restricted T cells may mediate cytotoxicity, DTH, and helper-inducer activity. (An alloreactive, class II MHC-restricted T-cell clone has shown all these activities – Dennert et al. 1981.)

Until recently, all Th cells tested were shown to be class II MHC restricted. Mizuochi et al. (1985) have now shown that both L3T4[+], Lyt2[−] and L3T4[−], Lyt2[+] cells may have Th activity and show synergy with *allospecific* Tc cell

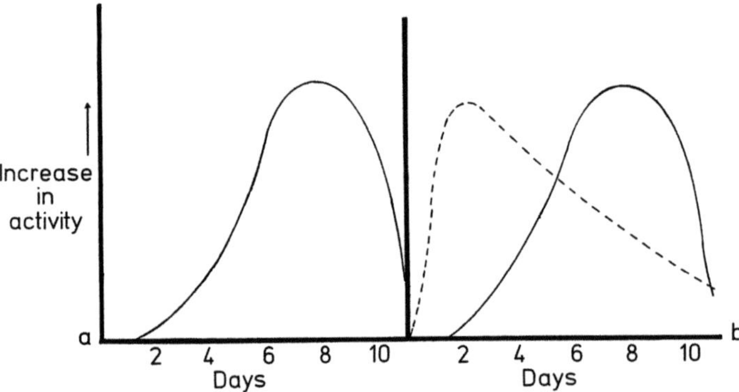

Fig. 1a, b. Approximate kinetics of appearance of T-cell activity in mouse spleens after one i.v. injection of influenza virus. **a** class I MHC restricted; ——, cytotoxic and DTH activity; **b** class II MHC restricted; -----, helper-inducer activity; ——, DTH activity. Data are taken from various publications referred to in ADA et al. (1981)

precursors, whereas *anti-TNP-Tc* cell responses were mediated only by L3T4$^+$, Lyt2$^-$ Th cells. In apparent contrast, SINICKAS et al. (1985) find that class I MHC-restricted Th cells could help Tc cell responses to murine cytomegalovirus infections (SINICKAS et al. 1985). Both classes of Th cells produced IL-2, highlighting the observation (LUKACHER 1985) that class I MHC-restricted cells which were inactive as Th cells did not secrete IL-2.

On this basis, it is not unreasonable to postulate that *during generation in a primary (in vivo) or secondary (in vitro) response*, cells of a given lineage may preferentially display different activities at different times while proliferating and differentiating after stimulation. The easiest way to study this is to measure T-cell responses in the spleen after a single i.v. injection of infectious virus – a single pulse, as only abortive replication occurs in mice. Figure 1 shows approximately the development of different activities over a 10-day period. In the case of class I MHC-restricted cells, very little Tc or Td activity is seen before day 4 and peak titers are reached between days 6 and 8; for both activities, Th are needed to obtain optimum activities (ASHMAN and MÜLLBACHER 1979; LEUNG and ADA 1981b). In the case of class II MHC-restricted cells, Th activity reaches peak titers at day 2 and decreases gradually thereafter, whereas Td activity reaches a peak at days 6–8. If these latter cells are of the one lineage and are self-sufficient for IL-2, other factors must determine when DTH activity is maximally shown. Morrison (quoted in LUKACHER et al. 1985) is reported to find H-2 I region-restricted killing in cultures of spleen cells shortly after in vitro stimulation with virus. It would be of interest to establish the profile of class II MHC-restricted Tc-cell activity in a primary response – does the activity peak at 2 or 7 days? Finally, it should be mentioned in this connection that, to date, it has been generally found that in primary or secondary responses to virus infections the great majority of Tc-cell activity is class I MHC antigen restricted. These findings may in part rest upon the widespread use of target cells which inadequately express certain class II MHC antigens. Though this aspect deserves further investigation, it has been widely found, using primary or secondary cell cultures, that transferred class I MHC-restricted cell populations are the more effective at giving protection. The demonstration of class II MHC-antigen-restricted influenza-specific Tc-cell clones during culture may simply indicate that more efficient production of IL-2 favors their emergence.

8.2 Regulation of the Response

Influenza viruses, especially those of the H2 and H6 subtype, are mitogens and the extent of the mitogenicity correlates with the presence of IE molecules on murine B lymphocytes (SCALZO and ANDERS 1984). Despite this, antibody production is T cell dependent as seen by a low antibody response in nu^+/nu^+ mice which returns to normal if thymocytes are adoptively transferred. Similarly, the production of antibody by human PBLs in vitro is T cell dependent. Similar results can be obtained with effector T cells, showing a basic requirement for T help in effector T-cell generation. However, there is little evidence that in normal mice the availability of Th cells is a limiting factor in the generation of a CMI response to influenza virus (e.g., LEUNG et al. 1982).

There is little information in this system on the formation of Ts cells and the role they may have in controlling the immune response. LIEW and RUSSELL (1980) showed that $Lyt1^+2^-$ cells which suppressed DTH activity could be detected in mouse spleen between 2 and 7 weeks after intranasal inoculation of infectious virus into mice.

From the point of view of making a successful vaccine to prevent influenza infection, it is important to know the response to different forms of the virus. There is general agreement that after parentera administration both infectious and inactivated virus induce antibody formation in mice. Similarly, Th and class II MHC-restricted Td cells are generated equally well (e.g., ADA et al. 1981). Several groups (e.g., BRACIALE and YAP 1978; WEBSTER and ASKONAS 1980) have shown that in mice, infectious virus is far more effective at inducing a class I MHC-restricted Tc-cell response than is inactivated virus. To some extent, the method of inactivation affects the result – virus which will undergo abortive infection (no viable progeny produced) may still be effective at inducing a Tc-cell response (e.g., ADA et al. 1981) and recently it has been found that influenza A virus, which was submitted to prolonged gamma-irradiation to destroy infectivity, not only could prime mice for a cross-reactive Tc-cell response but these mice were also protected against challenge by heterologous A strain viruses. This inactivation procedure was more effective than UV irradiation at "preserving" this property of the virus. The mechanism involved has not been elucidated (A. Müllbacher, G.L. Ada and R. Thattla, manuscript in preparation).

An in vitro Tc-cell response can be obtained from human PBLs by exposure to different virus preparations (MCMICHAEL et al. 1981; ENNIS et al. 1981). There is general agreement that whole inactivated virus adequately stimulates the response though the duration of the response has not been adequately determined. MCMICHAEL and colleagues found a subunit preparation less effective than did ENNIS and colleagues and this might be expected in view of the low proportion of cells reacting to the surface glycoproteins (Sect. 4). Responder T cells from human adults must be regarded as primed or memory cells and this probably explains the different responses to noninfectious virus between murine and human T cells.

8.3 Roles for Effector T Cells

Effector T cells or their precursors per se cannot prevent infection of susceptible cells by virus. The induction of effector T cells requires that viral antigen be presented by appropriate cells, such as macrophages or dendritic cells, or possibly by B cells. These cells should have "taken up" the virus by one or other mechanism or have been infected by the virus.

Effector T cells may contribute to two processes – to recovery from infection and/or to immunopathology which is part of the infectious process in normal hosts. CATE and MOLD (1975) found that if cells from the spleen and lymph nodes of mice taken 8–9 days after injection of formalinized, influenza A virus were transferred to syngeneic mice which were then challenged with an LD_{50} dose of virus, the recipient mice had a higher mortality rate than controls which received cells from nonimmunized mice. The increased mortality was prevented if the donor cells were pretreated with anti-Thy1 sera and C′ before transfer. If the donor mice had been immunized with infectious virus, cell transfer did not result in a higher mortality. The authors considered that their findings provided a rationale for the use of a live virus vaccine. A logical explanation was found for this result when YAP et al. (1978) showed that secondary effector T cells, obtained from the spleens of mice immunized with infectious virus and restimulated in vitro with infectious virus, were able to reduce lung virus titers when transferred to infected syngeneic mice. The cells responsible were $Lyt1^{-}2^{+}$, were class I MHC antigen restricted, and possessed Tc activity. Transfer of the same cell preparation to infected recipients which were class II MHC antigen compatible did not cause a reduction in lung virus titers. Similar cell preparations upon transfer to syngeneic recipients were found to cross-protect against different A strain viruses (YAP and ADA 1978). These findings were later confirmed (WELLS et al. 1983). LIN and ASKONAS (1981) and BRACIALE and colleagues (e.g., LUKACHER et al. 1984) subsequently showed that, when adoptively transferred, cloned T cells would reduce lung virus titers and protect mice from death. These findings agree with other related work. Thus, when nu^{+}/nu^{+} mice are infected with virus, the virus persists to high titer in the lungs for prolonged periods; transfer of immune spleen cells or secondary effector T cells to such mice rapidly reduces the level of virus in the lungs.

LUKACHER et al. (1984) showed that Tc-cell clones, which were either subtype specific or cross-reactive within A strains, could confer complete protection upon adoptive transfer to syngeneic mice infected with a lethal dose of influenza virus. Though the former was most likely specific for HA, the latter could have recognized one of several other virion components (Table 1), among which NP may be rather important. Andrew and colleagues (M. Andrew, G.L. Ada,

B. Coupar, and D. Boyle, manuscript in preparation), using secondary effector cell preparations with anti-NP Tc activity, and Taylor and Askonas (P.M. Taylor, B.A. Askonas, manuscript in preparation), using NP-specific Tc cell clones, have shown that adoptive transfer of these cells protects mice against lethal influenza infection.

Experiments with human volunteers also support the contention that Tc cells are important in the recovery from influenza virus infection. In studies at the Common Cold Centre in England, Tc cell memory (in PBLs) correlated with rapid clearing of administered virus in individuals, some of whom lacked specific antibody to HA or NA.

Following the early work of CATE and MOLD (1975), there is considerable evidence showing that the action of effector T cells induces pathological damage in the lungs, but it is not the only cause of such damage. Thus, YAP et al. (1979) showed that nu^+/nu^+ mice inoculated with about one LD_{50} dose of infectious virus displayed marked pathological lesions. On the other hand, a vigorous CMI response to an influenza virus infection can be mounted in the mouse lung with minimal pathology occurring, a deciding factor probably being the level of virus replication in the lung (MAK et al. 1982b). A limited amount of replication, occurring with a cold-adapted (ca) mutant virus, gave rise to a low level only of lung consolidation. If mice are treated with cyclosporin A (CsA) and infected with virus, the virus replicates to high titer but little immunopathology occurs because, although effector T cells develop, their activity is inhibited (SCHILTKNECHT and ADA 1985a), as is further discussed below. It was notable that if CsA was administered with a dose of virus lethal for normal mice, the mice survived and this was attributed to the greatly reduced level of pathological damage in the infected lung.

Adoptive transfer of class II MHC-restricted effector T cells with DTH but undetectable Tc activity has been shown by two groups (LEUNG and ADA 1982; LIEW and RUSSELL 1983) not to reduce lung virus titers but to reduce the survival of infected mice given a high dose of virus. Some evidence was obtained to show a linkage between this effect and the DTH activity of the transferred cells. These results may seem at odds with the demonstration by LUKACHER et al. (1985) that some clones of virus-specific class II MHC-restricted cells have Tc activity and when transferred can protect infected mice. At present, the findings of the two groups mentioned above, together with those of CATE and MOLD (1975), suggest that primary or secondary preparations of class II MHC-restricted cells either have a low proportion of cells with Tc activity and/or these cells may not be highly efficient at reducing lung virus titers, so that the effect of their presence is masked by the noncytotoxic cells.

8.4 Development and Mode of Action of Effector T Cells to Influenza Virus

The availability of agents such as cyclosporin A which inhibits the synthesis and/or secretion of lymphokine from cells (HODGKIN 1985) has made possible the study of factors which are important in the development and action of effector T cells. Experiments in the mid-1970s indicated that Tc cells in vitro acted directly, by cell-cell contact with target cells and not by the release and action of soluble factors at a distance. This is still believed to be valid. Thus, Tc cells lyse allo- or virus-infected target cells in the presence of CsA (ANDRUS and LAFFERTY 1982; SCHILTKNECHT and ADA 1985a). It became possible to analyze further the role of lymphokines in the development and mode of action of T cells during an influenza virus infection. If mice were treated with CsA and infected or injected with infectious virus, virus replication occurred and high virus titers persisted. Specific immune responses were delayed and of lesser magnitude (antibody, Tc-cell activity) or apparently absent (DTH activity – SCHILTKNECHT and ADA 1985a). But if cells from these mice or from untreated, infected mice were transferred to the hind footpad of untreated, syngeneic mice which had received an eliciting dose of virus 6 h previously, the two preparations of cells mediated an equivalent level of DTH, and this was shown to be class II MHC restricted (SCHILTKNECHT and ADA 1985b). That is, CsA prevented the action but not the formation of class II effector T cells but did substantially inhibit the formation of class I effector T cells. A possible explanation is that during generation, class II cells produce interleukins endogenously, whereas class I cells have a greater need of exogenous interleukins and CsA restricts the availability of these factors. Although some Tc-cell activity was detected in the lungs of CsA-treated, infected mice, their presence was detected in an in vitro test and this could not be taken as an indication of their in vivo activity. Therefore, secondary Tc cells were adoptively transferred to CsA-treated and untreated, infected mice. Experiments on the traffic of these cells showed that equal numbers reached the lungs in each group. In contrast to the controls, transfer of these cells failed to reduce lung virus titers in the CsA-treated mice (SCHILTKNECHT and ADA 1985c), suggesting a role for lymphokine release by Tc cells in vivo. The first suggestion of such a requirement was by TAYLOR and ASKONAS (1983), who described two influenza virus-specific T-cell clones. Only the clone which produced gamma interferon (IFN-γ) on contact with a target cell in vitro reduced lung virus titers in vivo.

ASKONAS and PALA (1985) went on to show that IFN-γ did not affect Tc-cell proliferation or maturation, but its release by Tc cells was entirely dependent on specific antigen recognition or mitogen treatment and correlated inversely with the growth rate of the clone and independently of cytotoxic activity at the different stages of maturation (TAYLOR et al. 1985). In contrast, LUKACHER et al. (1984), using two cytotoxic T-cell clones, one subtype specific and the other cross-reactive, showed that the induction and expression of antiviral effector activity by these two clones was highly specific, suggesting that the production of lymphokines by these Tc cells may not be critical. One way of resolving these apparent differences is to postulate (SCHILTKNECHT and ADA 1985c) that the lymphokines secreted by the effector T cells had an indirect role, namely, to enhance the expression of MHC antigens on the infected lung cells so they became more susceptible to T-cell lysis. It is known that IFN-γ is much more effective than IFN-$\alpha\beta$ at inducing MHC antigen expression in several cell types (e.g., WONG et al. 1983).

Lymphokine secretion is an important component of the DTH reaction, so the finding that effector T cells expressing this activity were present in the CsA-treated, infected mice but that their activity could not be elicited is not unexpected.

9 The Generation of Memory

9.1 B Cells

The persistence of serum antibody over decades and the occurrence of secondary antibody responses during successive influenza infections dictate that B-cell memory is long lasting. Specific B "memory" cells have been detected, following in vitro stimulation of peripheral blood lymphocytes (PBLs), in the majority of primed adults (CALLARD 1979). IgG-producing cells have been the predominant cell type detected; circulating IgA-producing cells have not been detected. The duration of B "memory" cells in the circulation after natural infection is unknown. After vaccination with an inactivated virus vaccine, antibody production from restimulated blood lymphocytes declined to prevaccination levels within 2 months, due to the disappearance of B memory cells from the circulation (MITCHELL et al. 1982). However, B memory cells have been detected in the spleen, lymph nodes, and tonsils of primed individuals in the absence of circulating B memory cells (CALLARD et al. 1982). In mice, HA-specific B-cell precursor frequency in the spleen increases 10- to 50-fold after primary infection (CANCRO et al. 1978). In addition, specific antibody-secreting cells generated in spleen cell cultures and detected by the hemolytic plaque assay increased tenfold after priming (McLAREN and POPE 1980).

Evidence for B-cell memory in the respiratory tract is suggested by the brisk IgA response in nasal secretions in primed children, after intranasal challenge with inactivated virus vaccine. Nasal wash IgA was not detectable by ELISA prior to viral challenge (WRIGHT et al. 1983). Furthermore, in vitro stimulation of human tonsillar tissue results in the production of specific IgA as well as IgG and IgM (McGAUGHAN et al. 1984).

9.2 T Cells

Memory for Tc-cell responses is the only aspect which has been examined in detail. One to 6 months after i.v. injection of virus, the frequency of Tc-cell precursors found in mouse spleens was at least tenfold higher than in spleens from unimmunized mice (ASKONAS et al. 1982b). The level of Tc precursor cells in mouse spleens 2 years after i.v. injection of virus was shown to be about half the level observed 3 weeks after i.v. injection of virus (ASHMAN 1982). Cells taken from the immunized mice at 2 years also showed Th-cell activity but this was not quantitated.

Possibly more relevant were studies on Tc-cell precursor frequency in mouse lung. Mice were inoculated intranasally with moderate doses of egg-grown influenza A virus, either of a parental strain or a cold-adapted (*ca-*) strain, with a mouse-adapted strain or with an UV-inactivated virus preparation. Infection with either of the egg-grown virus preparations caused a 20-fold increase in Tc-cell precursor frequency in the lungs of mice 2–6 weeks after virus administration, despite the fact that the parental strain replicated to much higher titers. With the mouse-adapted strain, precursor cell frequency was about 100-fold higher, whereas administration of inactivated virus had very little effect on precursor cell frequency (MAK et al. 1984). Though the possibility was not eliminated, little evidence was obtained for major traffic of Tc cells to or from the lungs during infection so the findings suggest that the precursor cells found in the lungs some time after viral replication are to a large extent progeny of the resident cells in normal lung.

Experiments in humans have been confined to PBL stimulation. Using PBLs from 189 human volunteers, MITCHELL et al. (1985) measured Tc-cell levels following stimulation with virus and culture for 6 days. Natural infection with influenza virus was shown to stimulate Tc-cell "memory" two- to fourfold. McMICHAEL et al. (1983) reported that there had been a low prevalence of influenza A infection in the Oxford and surrounding "catchment" areas since 1978, and they noticed that over a 5-year period there had been a sharp decline in the proportion of subjects examined who gave positive Tc-cell responses following in vitro stimulation of PBLs. They were led to conclude from this that the half-life of Tc memory cells in the circulation was 2–3 years. This is surprising in view of (1) the prolonged production of antibody to influenza virus (Sect. 9.1) and (2) the prolonged memory, sometimes said to be lifelong and in many cases 10 or more years, to a number of diseases such as smallpox and measles in which T-cell memory is probably an important component. It needs to be demonstrated in some way that sampling of PBLs is a reliable indication of the T-cell-immune status of an individual at different times.

10 Vaccination Against Influenza

10.1 Inactivated Virus Vaccines

Inactivated virus vaccines are prepared from the allantoic fluid from virus-infected eggs which is purified and concentrated by zonal centrifugation and inactivated. The main procedures used for inactivation of virus are treatment with formalin or β-propionolactone or UV light irradiation. Differences in the immunogenicity of inactivated virus vaccines, primarily the ability to prime the host for a Tc response, have been related to different methods of activation (ADA et al. 1981). Some experimental studies using inactivated virus vaccines have given incomplete details of procedures used for virus inactivation and this has led to some uncertainty as to whether inactivated virus vaccines can prime for a Tc response.

Whole-virus vaccines contain intact inactivated virus. Split-product vaccines are prepared from purified formalin-treated virus disrupted with chemicals to solubilize the viral envelope. The virus hemagglutinin and neuraminidase may be isolated and purified to produce subunit vaccines.

10.1.1 Immunogenicity

Sufficient data are now available to make an assessment of the immunogenicity of inactivated virus vaccines (POTTER 1982). In primed individuals, parenteral vaccination with either whole-virus or split-product vaccines results in a protective level of serum HI antibody in over 85% of recipients shortly after vaccination. In contrast, protective levels of serum HI antibody develop in approximately only 60% of unprimed recipients of whole-virus vaccines and to an even lesser extent after the split-product and subunit vaccines, requiring the administration of two doses of these vaccines to achieve an adequate response.

The induction of a secretory antibody response to inactivated virus vaccines is dependent both on the route of administration and on the recipient's prior antigenic experience. In unprimed subjects, local antibody responses (i.e., in nasal washes) are of low magnitude and occur infrequently after both parenteral and intranasal administration of vaccine. Parenteral administration produced a local IgG response, detected by ELISA, in 94% of primed recipients, whereas local IgA responses developed in only 38% (CLEMENTS et al. 1985). In contrast, local IgA responses develop in the majority of primed recipients after intranasal administration (WRIGHT et al. 1983). Although intranasal immunization with inactivated vaccine can induce protection against challenge virus infection, there are no comparative studies with parenteral administration. The oral administration of inactivated virus vaccine also induces a nasal wash antibody response in primed individuals, though data on efficacy are lacking (WALDMAN et al. 1981).

The duration of serum HI antibody after vaccination also varies according to the recipient's prior antigenic experience. Primed subjects retain protective levels of antibody for at least 1 year, whereas antibody levels decline rapidly in unprimed subjects.

The specificity of the antibody response after vaccination with inactivated virus in primed individuals is similar to that following natural infection (OXFORD et al. 1979), although a lesser degree of cross-reactivity to subtype variants is induced in persons initially immunized to a new subtype virus by inactivated vaccine rather than by natural infection (MASUREL et al. 1981).

The T-cell responses to inactivated virus vaccine have already been considered (Sect. 8.2). In primed humans, inactivated virus vaccines stimulate a cross-reactive Tc response (MCMICHAEL et al. 1981; ENNIS et al. 1981), whereas the ability of inactivated virus vaccine to stimulate the Tc response in unprimed humans has not been determined. In experimental studies in mice, virus inactivated by UV irradiation or formaldehyde induces a poor primary Tc response, possibly resulting from the generation of suppressor T cells, and is relatively ineffective in priming mice for a secondary cytotoxic response (ADA et al. 1981;

WEBSTER and ASKONAS 1980). In contrast, virus inactivated by gamma irradiation can prime mice for a cross-reactive Tc response as can a subunit vaccine containing only the nucleoprotein and hemagglutinin (WRAITH and ASKONAS 1985). In mice primed by infectious virus, inactivated virus induces secondary Tc and Td responses in vitro which are specific for the stimulating virus.

10.1.2 Efficacy

The protective efficacy of the different inactivated vaccines, including whole-virus and split-products, after parenteral administration are comparable, though the latter are less reactogenic, ranging from 60% to 80%.

In *primed* people (i.e., adults who have experienced one or more infections by influenza virus) immunized parenterally with inactivated vaccines, either whole- or split-virus preparations, and challenged with infectious homologous virus, vaccination may afford protection for several years. Though it is known that cross-reactive Tc cells are generated under these conditions, the extent to which they contribute to this result is not clear. In unprimed people, usually young children, this immunization is less protective, probably because of the poor ability of parenterally administered inactivated virus to *prime* for a local (respiratory tract) humoral or Tc-cell response. Even if given intranasally, inactivated virus does not *induce* a primary local humoral response in unprimed people. In contrast, a primary infection in children induces a local antibody response (MURPHY et al. 1982) and primes for a secondary IgA response (WRIGHT et al. 1983).

In view of antigenic drift, how effective is inactivated vaccine in protection against a challenge with a subtype variant virus? The study which best addresses this question is by HOSKINS et al. (1979), who showed that the protective effect of inactivated virus vaccine was limited to nonimmune schoolchildren who were vaccinated for the first time with the prevailing strain. Revaccination with later prevailing strains (inactivated) did not provide protection against subsequent challenge, whereas natural infection afforded almost complete protection during successive outbreaks involving drift viruses for more than 4 years. As the specificity of the antibody response after vaccination and infection is similar, differences in the extent of cross-reactive Tc responses to vaccination and infection may account for the lesser degree of heterotypic protection seen after vaccination. It would be of great interest to see a similar study over an equally substantial time period on adults.

10.2 Infectious Virus Vaccines

Influenza viruses have been effectively attenuated and the genes determining attenuation, which are located on the RNA segments coding for nonsurface antigens, transferred into reassortant viruses bearing the desired surface glycoproteins of current human influenza viruses. The possible approaches to producing attenuated reassortants have been recently reviewed (MURPHY and CHANOCK 1985).

10.2.1 Host Range Mutants

Attenuation of human influenza virus can be affected by the transfer of genes from host range mutants which are selected during repeated passage of human virus in eggs (e.g., A/PR/8/34(H1N1)). However, attenuation is not achieved if all of the six transferable PR-8 genes are transferred, as unexpectedly a mixture of the PR-8 and wild-type human influenza polymerases is required to restrict replication of the reassortant virus. Subsequently attenuation of reassortant viruses cannot be checked by simple in vitro methods, imposing a significant restriction on their use.

10.2.2 Temperature-Sensitive (ts) Mutants

ts mutants, produced by exposing virus-infected cells to a chemical mutagen (5-fluorouracil), are identified by being restricted in their ability to replicate in vitro at 37°–38° C. Although influenza A reassortant viruses bearing *ts* genes were satisfactorily attenuated, their genetic instability, leading to reversion to virulence, has precluded further use.

10.2.3 Cold-Adapted (ca) Mutants

ca mutants are produced by serial passage at successively lower temperatures (33°–25° C) in primary chick kidney cell culture. This causes mutations in each of the six transferable genes. The resulting *ca* mutant is also temperature sensitive and reassortant viruses receiving all six of the nonsurface antigen genes exhibit the *ca* and *ts* phenotypes. These reassortants are genetically stable (Cox and KENDAL 1984) and retain their *ca* phenotype even during infection in unprimed children who continue to shed virus for up to 12 days (BELSHE and VAN VORIS 1984). Furthermore, *ca* reassortant viruses were not transmitted to fully susceptible children who were exposed to vaccinated children (WRIGHT et al. 1982).

The human infectious dose 50 (HID_{50}), determined by antibody response and virus shedding, in seronegative adult volunteers (though presumably primed), is approximately $10^{5.5}$ to $10^{6.1}$ $TCID_{50}$ for H1N1 and H3N2 reassortant viruses (MURPHY and WEBSTER 1985). At a dose of $10^{7.5}$ $TCID_{50}$ or greater in adults, mild reactogenicity is observed. The HID_{50} for a H1N1 reassortant in unprimed children is 100-fold lower (BELSHE and VAN VORIS 1984). In children immunized with a comparable dose ($10^{6.3}$ $TCID_{50}$) of either an H1N1 or H3N2 reassortant, the H1N1 reassortant was less immunogenic than the H3N2 reassortant (WRIGHT et al. 1982). The dynamics of the serum and secretory antibody responses in primed and unprimed individuals have been previously described (Sect. 7.1.1).

Data on the T-cell response to live attenuated vaccines in humans are limited to the development of secondary Tc responses in primed individuals vaccinated with a PR-8 reassortant virus (ENNIS et al. 1981); there are no data on the

Tc responses in humans to *ca* reassortants. In mice, *ca* reassortants can induce a primary Tc response and can sensitize the lungs for a secondary Tc response (MAK et al. 1982; MAK et al. 1984). H3N2 reassortants are more effective than H1N1 reassortants at inducing a Tc response (TAO et al. 1985). The dose of a *ca* mutant of A/AA/6/60(H2N2) required to induce the same level of Tc response was 100–1000 times greater than that of the parental strain (MAK et al. 1982). A similar difference in dosage was required to prime mice to resist subsequent challenge with a mouse-adapted strain, A/WSN(H1N1). The difference in dosage required for priming could be overcome by giving two small doses 3 weeks apart (TANNOCK et al. 1984). Using this approach, *ca* reassortants may induce cross-protection against different subtype viruses, although the duration of cross-protection in mice is short (TANNOCK and PAUL 1985).

The protection afforded by *ca* reassortants has not been extensively evaluated. Seronegative adult volunteers were challenged with the homologous wild-type virus 1–2 months after vaccination with either an H3N2 *ca* reassortant ($10^{7.5}$ TCID$_{50}$) or inactivated vaccine (CLEMENTS et al. 1984). *Ca* recipients were completely protected against illness compared with a 72% efficacy in the inactivated virus vaccinees. Infection occurred in 19% of *ca* vaccinees and in 63% of inactivated virus vaccinees. Furthermore, nasal wash viral titers were 1000-fold lower in *ca* vaccinees shedding virus. Adult recipients of an H1N1 *ca* reassortant ($10^{7.8}$ TCID$_{50}$) were similarly protected against illness on challenge 1–3 months later (82% efficacy) and against infection – 18% of vaccinees were infected with 1000-fold lower viral titers than nonvaccinated controls (BETTS et al. 1985). At 6–7 months after vaccination, protective efficacy remained at 91% (CLEMENTS et al. 1985). Field trials on the protection afforded by *ca* reassortants against natural infection have not demonstrated superiority over inactivated virus vaccines in adults within the 1st year after vaccination, though *ca* reassortants were claimed to be more efficacious 2 years after vaccination against natural infection with a heterotypic H1N1 virus (COUCH et al. 1985).

The efficacy of *ca* reassortants has also been studied in unprimed children. Children receiving an H3N2 *ca* reassortant vaccine appeared to be protected against subsequent natural infection with related H3N2 strains (WRIGHT et al. 1982; BELSHE et al. 1984).

Ca reassortant vaccines are promising candidates for future influenza vaccines in view of their safety and immunogenicity, especially in unprimed children. Further studies are required to determine the duration of protection afforded and to determine if their efficacy in unprimed hosts and in heterotypic infection is superior to that of inactivated virus vaccines.

10.2.4 Avian Influenza Viruses

The genetic determinants of attenuation for avian viruses in primate cells reside on one or more of the genes coding for nonsurface antigens. These attenuating genes have been transferred into an avian-human reassortant which is restricted in its replication in primate cells. Infection with avian-human reassortant virus in monkeys induces significant resistance to subsequent challenge with wild-type human influenza virus. In humans, the reassortant was satisfactorily attenuated and immunogenic (MURPHY and WEBSTER 1985). However, the difficulty in confirming attenuation in vitro and the possible interaction with wild avian viruses may restrict this approach.

10.2.5 Deletion Mutants

Deletion mutants may be produced by treatment of DNA with restriction endonucleases. These mutants should be stable because of their inability to revert and low likelihood of suppression by a new mutation at another site on the viral genome. The potential for producing deletion mutants of influenza virus exists, as cloned DNA of the influenza genome has been inserted into SV40 vectors and the hemagglutinin expressed in eukaryotic cells. However, the transfer of genetic information in cloned DNA into a reassortant influenza virus has not been reported.

10.3 New Approaches to Vaccine Development

There are four main new approaches for the production of preparations which might form the basis of new vaccines. They are: (1) the synthesis of appropriate oligopeptides, (2) the synthesis in transformed prokaryotes or eukaryotes of foreign proteins, (3) the production of anti-idiotype antibodies, and (4) the use of infectious organisms as vectors of foreign DNA. Three of these approaches have been discussed earlier in this article in connection with antigen recognition.

Following the demonstration (GREEN et al. 1982) that antibodies could be formed against synthetic peptides representing a majority of the HA1 molecule and that such antibodies would bind to the HA molecule, the possibility of selecting a peptide from a conserved region of the HA1 which could be used as the basis of a vaccine was pursued by several groups. The results have generally been disappointing (Sect. 4). One group has reported the preparation of oligopeptides which might be used for the protective immunization against influenza infection. SHAPIRA et al. (1984) investigated regions that corresponded to two proposed antigenic sites. The largest peptide residues 138–164 which "encompassed" site A and part of site B, when coupled to tetanus toxoid, induced the formation of antibodies which reacted with the parent virus. Mice hyperimmunized with this preparation, but not with shorter peptides, were partially protected against challenge with the lowest dose of homologous virus which infected >90% of the mice. There remains a large gap between this result and what would be required to form an effective vaccine for human use, even against homologous virus.

DNA copies of all eight influenza virus genomic RNAs have been inserted into appropriate plasmid vectors and cloned in *E. coli* (quoted in PALESE and KINGSBURY 1983). These and some nonstructural viral antigens have been produced to high levels (e.g., YOUNG et al. 1983). With regard to the glycosylated proteins, the bacterial products have induced the formation of antibodies which were different from those elicited by the native protein or virus (NAYAK et al. 1984). This is not therefore an encouraging approach with glycosylated proteins and better results have been obtained with mammalian cells transfected with the cloned HA DNA gene (GETHING and SAMBROOK 1984; BRACIALE et al. 1984). The product of the transferred gene in the mammalian cell has the same chemical, physical, and antigenic properties as the viral HA and is recognized by Tc cells. So far, the use of this technology to produce viral proteins to be used for vaccination purposes has not proceeded very far.

A third approach is to use anti-idiotypes as surrogate antigens. This technology has certain advantages and disadvantages, among the latter being the problem of cost. Anti-idiotype preparations of an appropriate specificity have been prepared to a number of monoclonal antibodies and these have effectively prevented infection by the disease agent (e.g., ROITT et al. 1985) and they may induce both a humoral and a cell-mediated immune response (GELL and MOSS 1985; ERTL and FINBERG 1985). So far, there is no report of the preparation of an anti-idiotype antibody which prevents influenza infection in an experimental host.

A fourth approach involves the use of infectious agents, such as existing viral vaccines, as vectors of DNA coding for other protective antigens and there are a number of possibilities. Three viruses have considerable potential – vaccinia (SMITH et al. 1983; PANICALI and PAOLETTI 1982), herpes (ROIZMANN and JENKINS 1985), and adenovirus (DAVIS et al. 1985). Most work has been done with vaccinia virus. DNA coding for all influenza virus structural proteins and the three nonstructural proteins has been inserted into this virus (Table 1) and, where tested, most of these recombinant viruses have induced a primary antibody response and primed for a Tc cell response in mice. Recombinant vaccinia virus could be used as a vaccine against influenza virus, particularly as it is possible to insert quite large amounts of foreign DNA into the virus. Foreign DNA has also been inserted into herpes and adenoviruses. Recombinant adenovirus containing appropriate influenza DNA might be used to induce local immunity in the respiratory tract. Finally, mutant avirulent strains of *Salmonella* could also be used as a vector. Such recombinant bacteria would be administered orally and would induce primarily gut immunity; as there is considerable traffic of immune cells between mucosal sites in the body, significant immunological memory might be established in the respiratory tract after this form of immunization. These approaches are likely to be developed in the next few years.

11 Discussion and Conclusions

11.1 Contributions to Our Understanding of the Immune System

Many approaches have contributed to our understanding of the immune response, but from the beginning of the "Second Golden Age of Immunology" in the early 1960s, the use of viruses as tools to dissect the immune system – apart from epidemiological data – was a late starter. But from the early

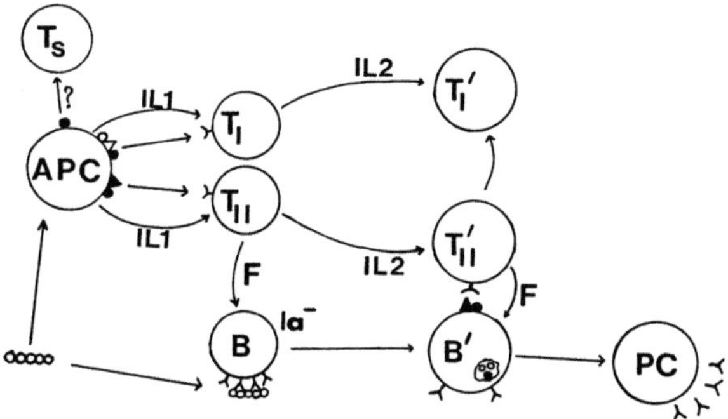

Fig. 2. Simplified diagrammatic representation of the immune system. Antigen (ooooo) is handled by the immune system in two ways: (1) By an antigen-presenting cell (*APC*) such as a macrophage or dendritic cell which processes the antigen (o, ●) and presents it in association with a class I (▵) or a class II (▴) MHC antigen to a precursor T cell, T_I and T_{II}, together with *IL1*. In the case of a virus, newly synthesized viral antigen may be presented in association with MHC antigens. The precursor T cells replicate and diffferentiate, $T_I \rightarrow T_I'$ (Tc, Td) and $T_{II} \rightarrow T_{II}'$ (Th, Td, Tc). (2) By a B cell which is selected by the specificity of the antigen. The antigen-receptor complexes are endocytosed and processed and antigenic fragments are presented in association with class II MHC antigens to an activated Th cell of the appropiate specificity. T-cell factors help to convert an Ia^- B cell to an Ia^+ cell and final differentiation to an antibody-secreting cell (adapted from ADA 1986). Induction of Ts activity by soluble antigen is also indicated, but this is a controversial topic

1970s, when the cell-mediated immune response could be studied more effectively, and the contributions which molecular biology, particularly sequencing, and X-ray crystallography became more apparent, the study of viruses and particularly influenza virus has progressively become more important. It is fair to claim now that the influenza virus is the most *comprehensively* studied of all viruses. Influenza virus studies have contributed to virtually every facet of the immune response as we understand it today and, in some respects, have provided data of critical importance. For example, the delineation of epitopes on the HA molecule which are involved in neutralization by antibody and the recognition by cytotoxic T cells of different viral proteins have not yet been matched by any other infectious agent.

It is generally thought that exposure to infectious agents must have had a profound influence on the development of the immune system and viruses were probably an important component of that array. Figure 2 is a diagrammatic representation of the immune system, as it might be seen by a virus particle which has polymeric protein structures. We postulate three main types of antigen-presenting cells – macrophages, dendritic cells, and B cells. Numerically, dendritic cells are very efficient (VAN VOORHIS et al. 1983) but, apart from their known role of presenting antigen to T cells, the mechanism of their action as antigen-presenting cells is at present unclear. Macrophages presumably process antigen in any form and particularly assembled structures, such as viruses and bacteria. Both these cells are thought to express the antigens, possibly as antigenic fragments at their surface in association with MHC antigens and these complexes are recognized by precursor T cells.

As far as class II MHC-restricted responses are concerned, it now seems very likely that the foreign antigenic material must have at least (and probably only) secondary structure in order to react with the MHC antigen, and this could be either as oligopeptides or as partially unfolded proteins. Whether one form is more effective than the other is not known. Similar information is not yet available for class I MHC-restricted responses. It may have been generally thought that, as the susceptibility of virus-infected target cells to lysis by effector cells required protein synthesis (e.g., JACKSON et al. 1976), intact viral protein molecules expressed at the cell surface became associated with the class I MHC molecule and this complex was recognized by the effector cell. This interpretation has now become less certain, especially as most internal viral proteins are now known to be recognized by the T cell. In view of the basic similarities in the structure of class I and class II MHC antigens, it could be thought that the form in which the foreign antigen was recognized might also have many similarities. This remains to be seen, but it can be anticipated that this will be a very active area for investigation in the next few years.

There is still much discussion about the nature of the T-cell receptor and although there has been a strong swing in recent years to the concept of a single recognition-single receptor model for T cells, a recent spirited commentary (LANGMAN and COHN 1985) maintains that the facts do not support this. We have not thought it necessary to review this field for this article. We look forward to the day when the isolated receptor is inserted into a synthetic membrane and the specificity of binding to other membrane-associated antigens, e.g., MHC gene products, is studied.

The third type of antigen-presenting cell is the B cell. The array of specific Ig receptors at the surface distinguish the B cell from the others. A major difference in antigen recognition was noted nearly 30 years ago when GELL and BENACERRAF (1959) showed that if ovalbumin (OVA)-immune guinea pigs were challenged, only native OVA could induce anaphylaxis (an antibody-mediated reaction), whereas both native and denatured OVA could induce DTH reactions. It became generally accepted that although antibody could be made to a great variety of substances, B cells more efficiently recognized conformational determinants (e.g., SELA 1969). High-resolution X-ray crystallography and the availability of monoclonal antibodies to viruses reinforces this belief in the case of influenza virus HA. Sites in the molecule which interact with neutralizing antibody are mainly discontinuous sequences. Though it may not yet be wise to eliminate the possibility that antigen is presented to B cells via another cell, e.g., the follicular dendritic cell, the B-cell receptors may directly bind free antigen in solution and the complex may be endocytosed and degraded. The recent elegant experiments of LANZAVECCHIA (1985) show how B cells may concentrate the antigen, process it, and present to the activated T cell an antigen-MHC complex with a specificity different from that of the Ig receptor on the B cell. This complex would mimic that on the surface of the macrophage/dendritic cell which initially selected the precursor of the T cell for activation. This work is an important extension to the Theory of Clonal Selection, explaining how antibody to a conformational determinant is selected and produced, the selecting antigen undergoing processing (resulting in destruction of the selecting determinant) after the selection process occurred.

Two developments might be expected: (1) That a proportion, possibly most, monoclonal antibodies which efficiently neutralize the infectivity of other viruses will recognize tertiary or quarternary conformations – usually discontinuous sequences; and (2) that in synthetic vaccine development, the use of peptides as B-cell determinants will involve large oligopeptides which retain the conformation of the parent protein to a considerable extent. The two peptides which are the closest to forming the basis of human peptide-based vaccines are either large (37 amino acids) or have a number of repeating sequences (quoted in ADA and SKEHEL 1985). A possible alternative is the synthesis a priori of peptides to bind to high titer to neutralizing monoclonal antibodies and their subsequent stabilization in that conformation (called a mimotope – GEYSON et al. 1986) so that they behave much like an anti-idiotype.

11.2 Influenza Vaccines for the Future

These considerations have implications for the design of future vaccines against influenza. Two conclusions from this review are: (1) In children, natural infection confers considerably better protection against subsequent subtype variants than does immunization with inactivated whole virus. This may also be the case for adults (who are already primed), but this has not been adequately documented. (2) We cannot yet predict antigenic changes in the HA molecule that occur in drift and even less so the changes that may occur in shift. Though every attempt should be made to predict such changes so that vaccination induces neutralizing antibody formation, the vaccine to be effective should also induce strong, cross-protective Tc-cell responses.

For now and the near future, there are four main choices: (1) A peptide-based vaccine. As our knowledge of the structural requirements for a peptide to bind to MHC antigens grows – assuming some further general principles emerge – it may be possible to synthesize a large oligopeptide which would contain sequences which bind with high affinity to Th, Tc, and B cells. A potential drawback to this approach is Ir gene effects. Analyses of the responses of a substantial number of individuals would need to be carried out to show whether a peptide-based vaccine would need to contain several peptides in each category in order to be effective for >90% of the population. (2) A subunit vaccine. The poor results of preparations based on HA:NA mixtures are probably due to their poor stimulation of cross-protective Tc cells. WRAITH and ASKONAS (1985) have proposed a mixture of HA and NP – the HA presumably to stimulate neutralizing antibody and the NP to induce cross-protective Tc-production. The concern about great variation in the murine response to NP (PALA et al. 1986) might be overcome if the DNA coding for these two antigens were incorporated into a live vector, such as vaccinia or *Salmonella*, or if these antigens were presented as ISCOMS – immunostimulating complexes (MOREIN et al. 1984). (3) Whole, inactivated virus. An inactivation process should be chosen which retains the ability of *all* components to induce good T-cell responses. For adults (i.e., already primed) this would seem to offer the best hope of a safe yet reasonably effective vaccine. At present, gamma-irradiation of the virus may be the method of choice for destroying infectivity. (4) Live attenuated virus. At present, *ca*-variant preparations are the most promising, but their efficacy needs to be established using a protocol similar to that used by HOSKINS et al. (1979). There is, however, concern about the widespread use of a live virus vaccine. A compromise would be to immunize children up to a given age with a live virus vaccine, as this would prime optimally for the most appropriate response (Sects. 8, 10). Inactivated whole-virus vaccine would be used after that age.

There are still some important findings to be made about influenza virus. It can be said that a vaccine is unlikely to be more efficient at inducing long-lasting immunity than infection by the wild-type virus. This can be as long as 20 years for a homologous strain challenge, as shown when H1N1 reemerged in 1977. This long-lasting immunity was almost certainly due to antibody. In contrast, immunity against antigen drift persists for 4–5 years. Is this due to the short half-life (about 3 years) of Tc-cell memory, as proposed by McMI-CHAEL et al. (1983)? If so, an important future aim is to see whether different immunization protocols can extend this period.

Acknowledgments. The authors wish to thank Drs. B.A. Askonas, J. Bennink, M. Andrew, and A. Müllbacher for permission to quote unpublished results.

Note Added in Proof

TOWNSEND et al. (Cell [to be published]) have recently shown that the epitopes of NP recognized by CTL in association with Class I molecules of the major histocompatibility complex in both mouse and man can be defined with short synthetic peptides derived from the NP sequence.

References

Ada GL (to be published) The generation of cellular versus humoral immunity. In: Arnon R (ed) Synthetic vaccines. CRC, Boca Raton

Ada GL, Skehel JJ (1985) Are peptides good antigens? Nature 316:764–765

Ada GL, Yap KL (1977) Matrix protein expressed at the surface of cells infected with influenza viruses. Immunochemistry 14:643–651

Ada GL, Leung K-N, Ertl HCJ (1981) An analysis of effector T cell generation and function in mice exposed to influenza A or Sendai viruses. Immun Rev 58:4–24

Allen PM, Matsueda GR, Haber E, Unanue ER (1985) Specificity of the T cell receptor: two different determinants are generated by the same peptide and the IA^k molecule. J Immunol 135:368–373

Anders EM, Katz JM, Jackson DC, White DO (1981) In vitro antibody response to influenza virus. II. Specificity of helper T cells recognising hemagglutinin. J Immunol 127:669–672

Andrew ME, Coupar BEH, Ada GL, Boyle DB (to be published) Cell-mediated immune responses to influenza virus antigens expressed by vaccinia virus recombinants. Microbial Pathogenesis

Andrewes CH, Bang FB, Burnet FM (1955) A short description of the myxovirus group (influenza and related viruses). Virology 1:176–184

Andrus L, Lafferty KJ (1982) Inhibition of T cell activity by cyclosporin A. Scand J Immunol 15:449–458

Ashman RB (1982) Persistence of cell-mediated immunity to influenza A virus in mice. Immunology 47:165–168

Ashman RB, Müllbacher A (1979) A helper T-cell for anti-viral cytotoxic T cell responses. J Exp Med 150:1277–1282

Askonas BA, Webster RG (1980) Monoclonal antibodies to the haemagglutinin and to H-2 inhibit the cross reactive T cell populations induced by influenza. Eur J Immunol 10:151–156

Askonas BA, Pala P (1985) Influenza specific cytotoxic T-cell clones and immune interferon. In: Feldman M, Lamb JR, Woody NJ (eds) Human T cell clones. Humana Press, New Jersey, pp 381–388

Askonas BA, McMichael AJ, Webster RG (1982a) The immune response to influenza viruses and the problem of protection against infection. In: Beare (ed) Basic and applied influenza research. CRC, Boca Raton, pp 159–188

Askonas BA, Müllbacher A, Ashman RB (1982b) Cytotoxic T memory cells in virus infection and the specificity of helper T cells. Immunology 45:79–84

Astry CL, Yolken RH, Jakab GJ (1984) Dynamics of viral growth, viral enzymatic activity and antigenicity in murine lungs during the course of influenza pneumonia. J Med Virol 14:81–90

Atassi MZ, Webster RG (1983) Localization, synthesis and activity of an antigenic site on influenza hemagglutinin. Proc Natl Acad Sci USA 80:840–844

Babbitt BP, Allen PM, Matsueda G, Haber E, Unanue ER (1985) Binding of immunogenic peptides to Ia histocompatibility molecules. Nature 317:359–361

Barry RD, Ives DR, Cruikshank JG (1962) Participation of deoxyribonucleic acid in the multiplication of influenza virus. Nature 194:1139–1140

Bean WJ (1984) Correlation of influenza A virus nucleoprotein with host species. Virology 133:438–442

Becht H, Huang RTC, Fleischer B, Boschek CB, Rott R (1984) Immunologic properties of the small chain HA2 of the haemagglutinin of influenza viruses. J Gen Virol 65:173–183

Belshe RB, Van Voris LP (1984) Cold-recombinant influenza A/California/10/76(H1N1) virus vaccine (CR-37) in seronegative children: infectivity and efficacy against investigational challenge. J Infect Dis 149:735–740

Belshe RB, Van Voris LP, Bartram J, Crookshanks FK (1984) Live attenuated influenza A virus vaccines in children: results of a field trial. J Infect Dis 150:834–840

Bennink JR, Yewdell JW, Gerhard W (1982) A viral polymerase involved in recognition of influenza virus infected cells by a cytotoxic T cell clone. Nature 296:75–76

Bennink JR, Yewdell JW, Smith GL, Moller C, Moss B (1984) Recombinant vaccinia virus primes and stimulates influenza haemagglutinin specific cytotoxic T cells. Nature 311:578–579

Betts RF, Douglas RG, Murphy BR (1985) Resistance to challenge with influenza A/Hong Kong/123/77(H1N1) wild-type virus induced by live attenuated A/Hong Kong/123/77(H1N1) cold-adapted reassortant virus. J Infect Dis 151:744–745

Biddison WE, Doherty PC, Webster RG (1977) Antibody to influenza virus matrix protein detects a common antigen on the surface of cells infected with type A influenza viruses. J Exp Med 146:690–697

Both GW, Sleigh MJ, Cox NJ, Kendal AP (1983) Antigenic drift in influenza H3 hemagglutinin from 1968–1980. Multiple evolutionary pathways and sequential amino acid changes at key antigenic sites. J Virol 48:52–60

Braciale TJ (1977) Immunologic recognition of influenza virus-infected cells. Cell Immunol 33:423–436

Braciale TJ (1979) Specificity of cytotoxic T cells directed to influenza virus haemagglutinin. J Exp Med 149:856–869

Braciale TJ, Yap KL (1978) Role of viral infectivity in the induction of influenza virus-specific cytotoxic T cells. J Exp Med 147:1236–1252

Braciale TJ, Andrew ME, Braciale VL (1981) Heterogeneity and specificity of cloned lines of influenza virus-specific cytotoxic T lymphocytes. J Exp Med 153:910–923

Braciale TJ, Braciale VL, Henkel TJ, Sambrook J, Gething MJ (1984) Cytotoxic T lymphocyte recognition of the influenza haemagglutinin gene product expressed by DNA mediated gene transfer. J Exp Med 159:341–354

Burlington DB, Clements ML, Meiklejohn G, Phelan M, Murphy BR (1983) Hemagglutinin specific antibody responses in immunoglobulin G, A and M isotypes as measured by ELISA after primary or secondary infection of humans with influenza A virus. Infect Immun 41:540–545

Burlington DB, Djeu JY, Wells MA, Kiley SC, Quinnan GV (1984) Large granular lymphocytes provide an accessory function in the in vitro development of influenza A virus-specific cytotoxic T cells. J Immunol 132:3154–3158

Callard RE (1979) Specific in vitro antibody response to influenza virus by human blood lymphocytes. Nature 282:734–736

Callard RE, McGaughan GW, Babbage J, Souhami RL (1982) Specific in vitro antibody responses by human blood lymphocytes: apparent nonresponsiveness of PBL is due to a lack of recirculating memory B cells. J Immunol 129:153–156

Cambridge G, MacKenzie JS, Keast D (1976) Cell-mediated immune response to influenza virus infection in mice. Infect Immun 13:36–43

Cancro MP, Gerhard W, Klinman NR (1978) The diversity of the influenza-specific primary B-cell repertoire in Balb/c mice. J Exp Med 147:776–787

Cate TR, Mold NG (1975) Increased influenza pneumonia mortality of mice adoptively immunized with node and spleen cells sensitized by inactivated but not live virus. Infect Immun 11:908–914

Cavanagh D, Mitkis F, Sweet C, Collie MH, Smith H (1979) The localisation of influenza virus in the respiratory tract of ferrets. J Gen Virol 44:505–514

Clements ML, O'Donnell S, Levine MM, Chanock RM, Murphy BR (1983) Dose response of A/Alaska/6/77(H3N2) cold-adapted reassortant vaccine virus in adult volunteers: role of local antibody in resistance to infection with vaccine virus. Infect Immun 40:1044–1051

Clements ML, Betts RF, Murphy BR (1984) Advantage of live attenuated cold-adapted influenza A virus over inactivated vaccine for A/Washington/80(H3N2) wild-type virus infection. Lancet i:705–708

Clements ML, Betts RF, Murphy BR (1985) Comparison of immune responses and efficacies of inactivated and live virus vaccines. J Cell Biochem Supp 9C:284–285

Colman PM, Varghese JN, Laver WG (1983) Structure of the catalytic and antigenic sites in influenza virus neuraminidase. Nature 303:41–44

Couch RB, Kasel JA (1983) Immunity to influenza in man. Ann Rev Microbiol 37:529–549

Couch RB, Kasel JA, Six HR, Cate TR (1981) The basis for immunity to influenza in man. In: Nayak (ed) Genetic variation among influenza viruses. Academic, New York, pp 535–546

Couch RB, Quarles JM, Cate TR, Zahradnik JM (1985) Field trials among college students with live attenuated influenza virus vaccines. J Cell Biochem Supp 9C:285

Cox NJ, Kendal AP (1984) Genetic stability of A/Ann Arbor/6/60 cold-mutant (temperature-sensitive) live influenza virus genes: analysis by oligonucleotide mapping of recombinant vaccine strains before and after replication in volunteers. J Infect Dis 149:194–200

Davis AR, Kostek B, Mason BB, Hsiao CL, Morin J, Dheer SK, Hung PP (1985) Expression of hepatitis B surface antigen with a recombinant adenovirus. In: Chanock P, Lerner R, Brown F (arranged) Modern approaches to vaccines. Cold Spring Harbor Laboratories, New York, p 85

Della-Porta AJ, Westaway EG (1978) A multihit model for the neutralization of animal viruses. J Gen Virol 38:1–19

DeLisi C, Berzofsky JA (1985) T cell antigenic sites tend to be amphipathic structures. Proc Natl Acad Sci USA 82:7048–7052

Dennert G, Weiss S, Warner JF (1981) T cells may express multiple activities: specific allohelp, cytolysis and delayed-type hypersensitivity are expressed by a cloned T-cell line. Proc Natl Acad Sci USA 78:4540–4543

Djeu JY, Stocks N, Zoon K, Stanton GJ, Tomonen T, Heberman RB (1982) Positive self regulation of cytotoxicity in human natural killer cells by production of interferon upon exposure to influenza and herpes viruses. J Exp Med 156:1222–1234

Doherty PC, Effros RB, Bennink J (1977) Heterogeneity of the cytotoxic response of thymus derived lymphocytes after immunization with influenza virus. Proc Natl Acad Sci USA 74:1209–1213

Doherty PC, Biddison WE, Bennink JR, Knowles BB (1978) Cytotoxic T cell responses in mice infected with influenza and vaccinia viruses vary in magnitude with H-2 genotype. J Exp Med 148:534–543

Effros RB, Frankel ME, Gerhard W, Doherty PC (1979) Inhibition of influenza immune T cell effector function by virus specific hybridoma antibody. J Immunol 123:1343–1346

Ennis FA, Meager A (1981) Immune interferon produced to high levels by antigenic stimulation of human lymphocytes with influenza virus. J Exp Med 154:1279–1289

Ennis FA, Meager A, Beare AS, Hua QY, Riley D, Schwarz G, Schild GC, Rook AH (1981) Interferon induction and increased natural killer-cell activity in influenza infections in man. Lancet 2:891–893

Ennis FA, Rook AH, Qi Y-L, Schild GC, Riley D, Pratt R, Potter CW (1981) HLA-restricted virus-specific cytotoxic T-lymphocyte responses to live and inactivated influenza vaccines. Lancet 2:887–891

Ertl HCJ, Finberg RW (1985) Induction of a virus specific immune response by an anti-idiotypic antibody. J Cell Biochem Supp 9C:294

Ettensohn DB, Roberts NJ (1984) Influenza virus infection of human alveolar and blood-derived macrophages: differences in accessory cell function and interferon production. J Infect Dis 149:942–949

Fischer A, Nash S, Beverley PCL, Feldmann M (1982) An influenza virus matrix protein-specific human T cell line with helper activity for in vitro anti-hemagglutinin antibody production. Eur J Immunol 12:844–849

Fleischer B, Becht H, Rott R (1985) Recognition of viral antigens by human influenza A virus specific T lymphocyte clones. J Immunol 135:2800–2804

Gangemi JD, Hightower JA, Jackson RA, Maker MH, Welsh MG, Sigel MM (1983) Enhancement of natural resistance to influenza virus in lipopolysaccharide-responsive and non-responsive mice by *Propionibacterium acnes*. Infect Immun 39:726–735

Gell PGH, Benacerraf B (1959) Studies on hypersensitivity. II. Delayed hypersensitivity to denatured proteins in guinea pigs. Immunology 2:64–70

Gell PGH, Moss PAH (1985) Production of cell-mediated immune response to *Herpes simplex* virus by immunization with anti-idiotype heteroantisera. J Gen Virol 66:1801–1804

Geraci JR, St Aubin DJ, Barker IK, Webster RG, Hinshaw VS, Bean WJ, Ruhnke HL, Prescott JH, Early G, Baker AS, Madoff S, Schooley RT (1982) Mass mortality of harbor seals. Pneumonia associated with influenza A virus. Science 215:1129–1131

Gerhardt W, Yewdell J, Frankel ME, Webster R (1981) Antigenic structure of influenza virus hae-magglutinin defined by hybridoma antibodies. Nature 290:713–716

Gething M-J, Sambrook J (1982) Construction of influenza hemagglutinin genes that code for intra-cellular and secreted forms of the protein. Nature 300:598–660

Geyson HM, Rodda SJ, Mason TJ (1986) The delineation of peptides able to mimic assembled epitopes. In: Synthetic peptides as antigens. Pitman, London, pp 130–149

Gonchoroff NJ, Kendal AP, Phillips DJ, Reimer CB (1982) Immunoglobulin M and G antibody response to type- and subtype-specific antigens after primary and secondary exposures of mice to influenza A viruses. Infect Immun 35:510–517

Green JA, Charette RP, Yeh T-H, Smith CB (1982) Presence of interferon in acute- and convalescent-phase sera of humans with influenza or an influenza-like illness of undetermined etiology. J Infect Dis 145:837–841

Green N, Alexander H, Olson A, Alexander S, Shinnick TM, Sutcliffe JG, Lerner RA (1982) Immuno-genic structure of the influenza virus hemagglutinin. Cell 28:477–487

Gresser I, Tovey MG, Maury C, Bandu MT (1976) Role of interferon in the pathogenesis of virus diseases in mice as demonstrated by the use of anti-interferon serum. II. Studies with Herpes simplex, Moloney sarcoma, vesicular stomatitis, Newcastle disease and influenza viruses. J Exp Med 144:1316–1323

Haaheim LR, Schild GC (1980) Antibodies to the strain-specific and cross-reactive determinants of the haemagglutinin of influenza H3N2 viruses. 2. Antiviral activities of the antibodies in biological systems. Acta Path Microbiol Scand B88:335–340

Hackett CJ, Askonas BA, Webster RG, Van Wyke K (1980) Quantitation of influenza virus antigens on infected target cells and their recognition by cross reactive cytotoxic T cells. J Exp Med 151:1014–1025

Hackett CJ, Dietzschold B, Gerhard W, Christ B, Knorr R, Gillessen D, Melchers F (1983) Influenza virus site recognised by a murine helper T cell specific for H1 strains. Localization to a nine amino acid sequence in the hemagglutinin molecule. J Exp Med 158:294–302

Hall WJ, Douglas RG (1980) Pulmonary function during and after common respiratory infections. Annu Rev Med 31:233–238

Haller O (1981) Inborn resistance of mice to orthomyxoviruses. Curr Top Microbiol Immunol 92:25–52

Hashimoto G, Wright PF, Karzon DT (1983) Antibody-dependent cell-mediated cytotoxicity against influenza virus-infected cells. J Infect Dis 148:785–794

Hay AJ, Lomniczi B, Bellamy AR, Skehel JJ (1977) Transcription of the influenza virus genome. Virology 83:337–355

Hicks JT, Ennis FA, Kim E, Verbonitz M (1978) The importance of an intact complement pathway in recovery from a primary viral infection: influenza in decomplemented and in C5-deficient mice. J Immunol 121:1437–1445

Hinshaw VS, Bean WJ, Webster RG, Easterday BC (1978) The prevalence of influenza viruses in swine and the antigenic and genetic relatedness of influenza viruses from man and swine. Virology 84:51–62

Hodgkin PD (1985) The activation and action of T lymphocytes. PhD Thesis, Australian National University, Canberra

Hoshino A, Takenaka H, Mizukoshi O, Imonishi J, Kishida T, Tovey MG (1983) Effect of anti-interferon serum on influenza virus infection in mice. Antiviral Res 3:59–65

Hoskins TW, Davies JK, Smith AJ, Miller CL, Allchin A (1979) Assessment of inactivated influenza-A vaccine after three outbreaks of influenza A at Christ's Hospital. Lancet 1:33–35

Hruskova J, Syrucek L, Tumova B, Stumpa A, Bruckova M, Losova M, Berkovicova V (1976) Levels of immunoglobulins and antibodies to haemagglutinin and neuraminidase of influenza virus in nasal secretions after natural infection. Acta Virol 30:126–134

Hurwitz JL, Heber-Katz E, Hackett CJ, Gerhard WJ (1984) Characterization of the murine T_H response to influenza virus hemagglutinin. Evidence for three major specificities. J Immunol 133:3371–3377

Husseini RH, Sweet C, Collie MH, Smith H (1981) The relation of interferon and nonspecific inhibitors to virus levels in nasal washes of ferrets infected with influenza viruses of differing virulence. Br J Exp Pathol 62:87–93

Husseini RH, Sweet C, Collie MH, Smith H (1982) Elevation of nasal viral levels by suppression of fever in ferrets infected with influenza viruses of differing virulence. J Infect Dis 145:520–524

Husseini RH, Sweet C, Overton H, Smith H (1984) Role of maternal immunity in the protection of newborn ferrets against infection with a virulent influenza virus. Immunology 52:389–394

Jackson DC, Ada GL, Hapel AJ, Dunlop MBC (1976) Changes in the surface of virus-infected cells recognised by cytotoxic T cells. II. A requirement for glycoprotein synthesis in virus-infected target cells. Scand J Immunol 5:1021–1029

Jackson DC, Murray JM, White DO, Fagan CN, Tregear GW (1982) Antigenic activity of a synthetic peptide comprising the "loop" region of influenza virus hemagglutinin. Virology 120:273–276

Kaplan DV, Braciale VL, Braciale TJ (1984) Antigen-dependent regulation of IL-2 receptor expression of cloned human cytotoxic T lymphocytes. J Immunol 133:1966–1969

Katz JM, Laver WG, White DO, Anders EM (1985) Recognition of influenza virus hemagglutinin by sub-type specific and cross reactive proliferative T cells; contribution of HA1 and HA2 polypeptide chains. J Immunol 134:616–622

Kees U, Krammer PH (1984) Most influenza A virus specific memory cytotoxic T lymphocytes react with antigenic epitopes associated with internal virus determinants. J Exp Med 159:365–377

Kosinowski VH, Allen H, Gething MJ, Waterfield MD, Klenk H (1980) Recognition of viral glycoproteins by influenza A specific cross reactive cytotoxic T lymphocytes. J Exp Med 151:945–958

Kris RM, Asofsky R, Evans CB, Small PA (1985) Protection and recovery in influenza virus-infected mice immunosuppressed with anti-IgM. J Immunol 132:1230–1235

Lamb RA, Choppin PW (1983) Gene structure and replication of influenza virus. Virology 81:382–397

Lamb JR, Feldmann M (1982) A human suppressor T cell clone which recognises an autologous helper T cell clone. Nature 300:456–468

Lamb JR, Eckels DD, Lake P, Woody JN, Green N (1982) Human T cell clones recognise chemically synthesized peptides of influenza hemagglutinin. Nature 300:66–69

Lamb JR, Skidmore BJ, Green N, Chiller JM, Feldmann M (1983) Induction of tolerance in influenza virus-immune T-lymphocyte clones with synthetic peptides of influenza hemagglutinin. J Exp Med 157:1434–1447

Lambre CR, Thibon M, Le Maho S, Di Bella G (1983) Autoantibody dependent activation of the autologous classical complement pathway by guinea-pig red cells treated with influenza virus or neuraminidase. Immunology 49:311–319

Langman RE, Cohn M (1985) T cell function via restricted recognition of antigen, not antigen-restricted recognition. Cell Immunol 94:598–608

Lanzavecchia A (1985) Antigen-specific interaction between T and B cells. Nature 314:537–539

Laver WG, Valentine RC (1969) Morphology of the isolated hemagglutinin and neuraminidase subunits of the influenza virus. Virology 38:105–119

Lee PWK, Hayes EC, Joklik WC (1981) Protein 61 is the reovirus attachment protein. Virology 108:156–163

Leung KN, Ada GL (1981a) Induction of natural killer cells during murine influenza virus infection. Immunobiol 160:342–366

Leung KN, Ada GL (1981b) The effect of helper T cells on the primary in vitro production of delayed-type hypersensitivity to influenza virus. J Exp Med 153:1029–1043

Leung K-N, Ada GL (1982) Different functions of subsets of effector T cells in murine influenza virus infection. Cell Immunol 67:312–324

Leung K-N, Schiltknecht E, Ada GL (1982) In vivo collaboration between precursor T cells and helper T cells in the development of delayed-type hypersensitivity reaction to influenza virus in mice. Scand J Immunol 16:257–264

Liew FY, Russell SM (1980) Delayed-type hypersensitivity to influenza virus. J Exp Med 151:799–814

Liew FY, Russell SM (1983) Inhibition of pathogenic effect of effector T cells by specific suppressor T cells during influenza infection in mice. Nature 304:541–543

Lin YL, Askonas BA (1981) Biological properties of an influenza A virus-specific killer T cell clone. J Exp Med 154:225–234

Lukacher AE, Braciale VL, Braciale TJ (1984) In vivo effector function of influenza virus-specific cytotoxic T lymphocyte clones is highly specific. J Exp Med 160:814–826

Lukacher AE, Morrison LA, Braciale VL, Malissen B, Braciale TJ (1985) Expression of specific cytolytic activity by H-2 I region-restricted, influenza specific T lymphocyte clones. J Exp Med 162:171–187

Lyons CR, Lipscomb MF (1983) Alveolar macrophages in pulmonary immune responses. I. Role in the initiation of primary immune responses and in the selective recruitment of T lymphocytes to the lung. J Immunol 130:1113–1119

Mahy BWJ, Hastie ND, Armstrong SJ (1972) Inhibition of influenza virus replication by alpha-amantin: mode of action. Proc Natl Acad Sci USA 69:1421–1424

Mak NK, Leung KN, Ada GL (1982a) The generation of "cytotoxic" macrophages in mice during infection with influenza A or Sendai virus. Scand J Immunol 15:553–561

Mak NK, Zhang Y-L, Ada GL, Tannock GA (1982b) Humoral and cellular responses of mice to infection with a cold adapted influenza A virus variant. Infect Immun 38:218–225

Mak NK, Schiltknecht E, Ada GL (1983) Protection of mice against influenza virus infection: enhancement of nonspecific cellular responses by Corynebacterium parvum. Cell Immunol 78:314–325

Mak NK, Sweet C, Ada GL, Tannock GA (1984) The sensitization of mice with a wild-type and cold-adapted variant of influenza A virus. II. Secondary cytotoxic T cell responses. Immunology 51:407–416

Manca F, Clarke JA, Miller A, Sercarz EE, Shastri N (1984) A limited region within hen egg-white lysozyme serves as the focus for a diversity of T cell clones. J Immunol 133:2075–2078

Masurel N, Ophof P, de Jong P (1981) Antibody response to immunisation with influenza A/USSR/71(H1N1) virus in young individuals primed or unprimed for A/New Jersey/76(H1N1) virus. J Hyg Camb 87:201–209

McDermott MR, Befus AD, Bienenstock J (1982) The structural basis for immunity in the respiratory tract. Int Rev Exp Pathol 23:47–112

McGaughan GW, Adams E, Basten A (1984) Human antigen-specific IgA responses in blood and secondary lymphoid tissue: an analysis of help and suppression. J Immunol 132:1190–1196

McLaren C, Butchko GM (1978) Regional T- and B-cell responses in influenza-infected ferrets. Infect Immun 22:189–194

McLaren C, Pope B (1980) Macrophage dependency of in vitro B cell response to influenza virus antigens. J Immunol 125:2679–2684

McMichael AJ, Ting A, Zweerink HJ, Askonas BA (1977) HLA restriction of cell mediated lysis of influenza virus-infected human cells. Nature 270:524–526

McMichael AJ, Gotch F, Eullen P, Askonas BA, Webster RG (1981) The human cytotoxic T cell response to influenza A vaccination. Clin Exp Immunol 43:267–284

McMichael AJ, Gotch FM, Dongworth DW, Clark A, Potter CW (1983) Declining T-cell immunity to influenza 1977–1982. Lancet 2:762–764

McMichael AJ, Michie CA, Gotch FM, Smith GL, Moss B (to be published) Recognition of influenza A virus nucleoprotein by human cytotoxic T lymphocytes. J Gen Virol

Milich DR, Thornton GB, McLachlan A, McNamara MK, Chisari FV (to be published) T and B cell recognition of native and synthetic pre-S-region determinants of HBsAg. In: Chanock RM, Lerner RA, Brown F (eds) Modern approaches to vaccines. Cold Spring Harbor Laboratory, New York

Mitchell DM, Fitzharris P, Knight RA, Schild GC (1982) Kinetics of specific in vitro antibody production following influenza immunisation. Clin Exp Immunol 48:491–498

Mitchell DM, McMichael AJ, Lamb JR (1985) The immunology of influenza. Br Med Bull 41:80–85

Mizuochi T, Golding H, Rosenberg AS, Glimcher LH, Malek TK, Singer A (1985) Both L3T4$^+$ and Lyt2$^+$ helper T cells initiate cytotoxic T lymphocyte responses against allogeneic major histocompatibility antigens but not against trinitrophenyl-modified self. J Exp Med 162:427–443

Morein B, Sundquist B, Höglund S, Dalsgaard K, Osterhaus A (1984) Iscom, a novel structure for antigenic presentation of membrane proteins from enveloped viruses. Nature 308:457–560

Muller GM, Shapira M, Arnon R (1982) Anti-influenza response achieved by immunization with a synthetic conjugate. Proc Natl Acad Sci USA 79:569–573

Murphy BR, Chanock RM (1985) Immunization against viruses. In: Fields BN (ed) Virology. Raven, New York, pp 349–370

Murphy BR, Webster RG (1985) Influenza viruses. In: Fields BN, Knipe DM, Chanock RM, Melnick JL, Roizman B, Shope RE (eds) Virology. Raven, New York, pp 1179–1239

Murphy BR, Baron S, Chalhub EG, Uhlendorf CP, Chanock RM (1973) Temperature sensitive mutants of influenza virus. IV. Induction of interferon in the nasopharynx by wild-type and a temperature-sensitive recombinant virus. J Infect Dis 128:488–493

Murphy BR, Nelson DL, Wright PF, Tierney EL, Phelan MA, Chanock RM (1982) Secretory and systemic immunological response in children infected with live attenuated influenza A virus vaccines. Infect Immun 36:1102–1108

Nayak DP, Davis AR, Ueda M, Bos TJ, Sivasubramanian N (1984) Characterization of influenza virus glycoproteins expressed from cloned cDNAs in prokaryotic and eukaryotic cells. In: Chanock RM, Lerner RA (eds) Modern approaches to vaccines. Cold Spring Harbor Laboratory, New York, pp 165–172

Nestorowicz A, Laver G, Jackson DC (1985a) Antigenic determinants of influenza virus haemagglutinin. X. A comparison of the physical and antigenic properties of monomeric and trimeric forms. J Gen Virol 65:1687–1695

Nestorowicz A, Tregear GW, Southwell CN, Martyn J, Murray JM, White DO, Jackson DC (1985b) Antibodies elicited by influenza virus hemagglutinin fail to bind to synthetic peptides representing putative antigenic sites. Mol Immunol 22:145–154

Owen JA, Allouche M, Doherty PC (1982) Limiting dilution analysis of the specificity of influenza-immune cytotoxic T cells. Cell Immunol 67:49–59

Oxford JS, Schild GC, Potter CW, Jennings R (1979) The specificity of the anti-haemagglutinin antibody response induced in man by inactivated influenza vaccines and by natural infection. J Hyg Camb 82:51–61

Oxford JS, Haaheim LR, Slepushkin A, Werner J, Kuwert E, Schild GC (1981) Strain specificity of serum antibody to the haemagglutinin of influenza A (H3N2) viruses in children following immunisation or natural infection. J Hyg Camb 86:17–26

Pala P, Askonas BA (1985) Induction of K^b-restricted anti-influenza cytotoxic T cells in C57Bl mice: importance of stimulator cell type and immunization route. Immunology 55:601–607

Pala P, Townsend ARM, Askonas BA (1986) Viral recognition by influenza A virus crossreactive cytotoxic T cells: the proportion of Tc that recognise nucleoprotein varies between individual mice. Eur J Immunol 16:193–198

Palese P, Kingsbury DW (eds) (1983) The genetics of influenza viruses. Springer, Vienna New York

Palese P, Ritchey MB, Schulman JL (1977) P1 and P3 proteins of influenza virus are required for complementary RNA syntheses. J Virol 21:1187–1195

Panicali D, Paoletti E (1982) Construction of poxviruses as cloning vectors: insertion of the thymidine kinase gene from herpes simplex virus into the DNA of infectious vaccinia virus. Proc Natl Acad Sci USA 79:4927–4931

Perrin LH, Joseph BS, Cooper NR, Oldstone MBA (1976) Mechanism of injury of virus-infected cells by antiviral antibody and complement: participation of IgG, F(ab′)₂, and the alternative complement pathway. J Exp Med 143:1027–1041

Possee RD, Schild GC, Dimmock NJ (1982) Studies on the mechanisms of neutralization of influenza virus by antibody: evidence that neutralizing antibody (antihaemagglutinin) inactivates influenza virus in vivo by inhibiting virion transcriptase activity. J Gen Virol 58:373–386

Potter CW (1982) Inactivated influenza virus vaccine. In: Beare AS (ed) Basic and applied influenza research. CRC Press, Florida, pp 119–156

Puck JM, Glezen WP, Frank AL, Six HR (1980) Protection of infants from infection with influenza A virus by transplacentally acquired antibody. J Infect Dis 142:844–849

Reiss CS, Schulman JL (1980a) Influenza type A virus M protein expression on infected cells is responsible for cross-reactive recognition by cytotoxic thymus-derived lymphocytes. Infect Immun 29:719–723

Reiss CS, Schulman JL (1980b) Cellular immune responses of mice to influenza virus infection. Cell Immunol 56:502–509

Reiss CS, Burakoff SJ (1981) Specificity of the helper T cell for the cytolytic T lymphocyte response to influenza viruses. J Exp Med 154:541–546

Reuman PD, Paganini CMA, Ayoub EM, Small PA (1983) Maternal-infant transfer of influenza-specific immunity in the mouse. J Immunol 130:932–936

Rodgers BC, Mims CA (1981) Interaction of influenza virus with mouse macrophages. Infect Immun 31:751–757

Rodgers BC, Mims CA (1982a) Influenza virus replication in human alveolar macrophages. J Med Virol 9:177–184

Rodgers BC, Mims CA (1982b) Role of macrophage activation and interferon in the resistance of alveolar macrophages from infected mice to influenza virus. Infect Immun 36:1154–1159

Roitt IM, Thanavala YM, Male DK, Hay FC (1985) Anti-idiotypes as surrogate antigens: structural considerations. Immunol Today 6:265–267

Roizman B (1985) Multiplication of viruses: an overview. In: Fields BN et al. (eds) Virology. Raven, New York, pp 69–75

Roizman B, Jenkins FJ (1985) Genetic engineering of novel genomes of large DNA viruses. Science 229:1208–1213

Rosenthal AS, Shevach EM (1973) Function of macrophages in antigen recognition by guinea pig T lymphocytes. J Exp Med 138:1194–1212

Rosztoczy I, Sweet C, Toms GL, Smith H (1975) Replication of influenza virus in organ cultures of human and simian urogenital tissues and human foetal tissues. Br J Exp Pathol 56:322–328

Russell SM, Liew FY (1979) T cells primed by influenza virion internal components can co-operate in the antibody response to haemagglutinin. Nature 280:147–148

Scalzo AA, Anders EM (1984) Role of I-E antigens in the mitogenic response to influenza viruses. Proc Aust Soc Imm 14:93

Schiltknecht E, Ada GL (1985a) In vivo effects of cyclosporin on influenza A virus-infected mice. Cell Immunol 91:227–239

Schiltknecht E, Ada GL (1985b) The generation of effector T cells in influenza A-infected, cyclosporine A-treated mice. Cell Immunol 95:340–348

Schiltknecht E, Ada GL (1985c) Influenza virus-specific T cells fail to reduce lung virus titers in cyclosporin-treated, infected mice. Scand J Immunol 22:99–103

Schulman JL (1975) Immunology of influenza. In: Kilbourne ED (ed) Influenza viruses and influenza. Academic, New York, pp 373–393

Sela M (1969) Antigenicity: some molecular aspects. Science 166:1365–1374

Shapira M, Jibson M, Muller G, Arnon R (1984) Immunity and protection against influenza virus by synthetic peptide corresponding to antigenic sites of hemagglutinin. Proc Natl Acad Sci USA 81:2461–2465

Shaw MW, Lamon EW, Compans RW (1981) Surface expression of a non-structural antigen on influenza A virus-infected cells. Infect Immun 34:1065–1068

Shaw MW, Lamon EW, Compans RW (1982) Immunologic studies of the influenza A virus non-structural protein NS1. J Exp Med 156:243–254

Shvartsman YS, Zykov MP (1976) Secretory anti-influenza immunity. Adv Immunol 22:291–330
Shvartsman YS, Agranovskaya EN, Zykov MP (1977) Formation of secretory and circulating antibodies after immunisation with live and inactivated influenza virus vaccines. J Infect Dis 135:697–705
Simons K, Garoff H, Helenius A (1982) How an animal virus gets into and out of its host cell. Sci Am 246:58–66
Sinickas VG, Ashman RB, Hodgkin PD, Blanden RV (1985) The cytotoxic response to murine cytomegalovirus. III. Lymphokine release and cytotoxicity are dependent upon phenotypically similar immune cell populations. J Gen Virol 66:2551–2562
Skehel JJ, Bayley PM, Brown EB, Martin SR, Waterfield DM, White JM, Wilson IA, Wiley DC (1982) Changes in the conformation of influenza virus hemagglutinin at the pH optimum of virus-mediated membrane fusion. Proc Natl Acad Sci USA 79:968–972
Smith GL, Murphy BR, Moss B (1983) Construction and characterisation of an infectious vaccinia virus recombinant that expresses the influenza haemagglutinin gene and induces resistance to influenza virus infection in hamsters. Proc Natl Acad Sci USA 80:7155–7159
Stein-Streilein J, Bennett M, Mann D, Kumor V (1983) Natural killer cells in mouse lung: surface phenotype, target preference and response to local influenza virus infection. J Immunol 131:2699–2704
Sterkers G, Michon J, Henin Y, Gomard E, Hannoun C, Levy JP (1985) Fine specificity analysis of human influenza-specific cloned cell lines. Cell Immunol 94:394–405
Stitz L, Huang RTC, Hengartner H, Rott H, Zinkernagel RM (1985) Cytotoxic T cell lysis of target cells fused with liposomes containing influenza virus haemagglutinin and neuraminidase. J Gen Virol 66:1333–1339
Streicher HZ, Berkower IJ, Busch M, Gurd FRN, Berzofsky JA (1984) Antigen conformation determines processing requirements for T-cell activation. Proc Natl Acad Sci USA 81:6831–6835
Sweet C, Smith H (1980) Pathogenicity of influenza virus. Microbiol Rev 44:303–330
Sweet C, Cavanagh D, Collie MH, Smith H (1978) Sensitivity to pyrexial temperatures: a factor contributing to virulence differences between two clones of influenza virus. Br J Exp Pathol 59:373–380
Sweet C, Macartney JC, Bird RA, Cavanagh D, Collie MH, Husseini RH, Smith H (1981) Differential distribution of virus and histological damage on the lower respiratory tract of ferrets infected with influenza viruses of differing virulence. J Gen Virol 54:103–114
Tannock GA, Paul GA (1985) Homologous and heterologous immunity to influenza A virus induced by recombinants of the cold-adapted master strain A/Ann Arbor/6/60-Ca. J Cell Biochem Supp 9C:294
Tannock GA, Paul JA, Barry RD (1984) Relative immunogenicity of the cold-adapted influenza virus A/Ann Arbor/6/60 (A/AA/6/60-ca), recombinants of A/AA/6/60-ca and parental strains with similar surface antigens. Infect Immun 43:457–462
Tao SJ, Mak NK, Ada GL (1985) Sensitization of mice with wild-type and cold-adapted influenza virus variants: immune responses to two H1N1 and H3N2 viruses. J Virol 53:645–650
Taylor HP, Dimmock NJ (1985a) Mechanisms of neutralization of influenza virus by secretory IgA is different from that of monomeric IgA or IgG. J Exp Med 161:198–209
Taylor HP, Dimmock NJ (1985b) Mechanisms of neutralization of influenza virus by IgM. J Gen Virol 66:903–907
Taylor PA, Askonas BA (1983) Diversity in the biological properties of anti-influenza cytotoxic T cell clones. Eur J Immunol 13:707–711
Taylor PM, Wraith DC, Askonas BA (1985) Control of immune interferon release by cytotoxic T-cell clones specific for influenza. Immunology 54:607–614
Thomas DB, Hackett CJ, Askonas BA (1982) Evidence for two T-helper populations with distinct specificity in the humoral response to influenza A viruses. Immunology 47:429–436
Toms GL, Davies JA, Woodward CG, Sweet C, Smith H (1977) The relation of pyrexia and nasal inflammatory response to virus levels in nasal washings of ferrets infected with influenza viruses of differing virulence. Br J Exp Pathol 58:444–458
Townsend ARM, McMichael AJ (1985) Specificity of cytotoxic T lymphocytes stimulated with influenza virus. Prog Allergy 36:10–43
Townsend ARM, Skehel JJ (1984) The influenza virus nucleoprotein gene controls the induction of both subtype specific and cross reactive T cells. J Exp Med 160:552–563

Townsend ARM, McMichael AJ, Carter NP, Huddleston JA, Brownlee GG (1984a) Cytotoxic T cell recognition of the influenza nucleoprotein and haemagglutinin expressed in transfected mouse L cells. Cell 39:13–25

Townsend ARM, Skehel JJ, Taylor PM, Palese P (1984b) Recognition of influenza A virus nucleoprotein by an H-2 restricted cytotoxic T cell clone. Virology 133:456–459

Townsend ARM, Gotch FM, Davey J (1985) Cytotoxic T cells recognise fragments of the influenza nucleoprotein. Cell 42:457–567

Van Voorhis WC, Valnisky J, Hoffman E, Luban J, Hair LS, Steinman RN (1983) The relative efficacy of human monocytes and dendritic cells as accessory cells for T cell regulation. J Exp Med 158:174–191

Van Wyke KL, Hinshaw VS, Bean WJ, Webster RG (1980) Antigenic variation of influenza A virus nucleoprotein detected with monoclonal antibodies. J Virol 35:24–30

Van Wyke KL, Yewdell JW, Reck LJ, Murphy BR (1984) Antigenic characterization of influenza A virus matrix protein with monoclonal antibodies. J Virol 49:248–252

Varghese JN, Laver WG, Colman PM (1983) Structure of the influenza virus glycoprotein antigen neuraminidase at 2.9f resolution. Nature 303:35–40

Verbonitz MW, Ennis FA, Hicks JT, Albrecht P (1978) Hemagglutinin-specific complement-dependent cytolytic antibody response to influenza infection. J Exp Med 147:265–270

Virelizier JL (1975) Host defenses against influenza virus: the role of anti-hemagglutinin antibody. J Immunol 115:434–439

Virelizier JL, Oxford JS, Schild GC (1976) The role of humoral immunity in host defence against influenza A infection in mice. Postgrad Med J 52:332–337

Virelizier JL, Allison A, Oxford C, Schild GC (1977) Early presence of nucleoprotein antigen on the surface of influenza virus-infected cells. Nature 266:52–53

Wabuke-Bunoti MAN, Fan DP (1983) Isolation and characterization of a CNBr cleavage peptide of influenza viral hemagglutinin stimulatory for mouse cytolytic T lymphocytes. J Immunol 130:2386–2391

Waldman RH, Wood SH, Torres EJ, Small PA (1970a) Influenza antibody response following aerosol administration of inactivated virus. Am J Epidem 91:575–584

Waldman RH, Wigley FM, Small PA (1970b) Specificity of respiratory secretion antibody against influenza virus. J Immunol 104:1477–1483

Waldman RH, Jurgensen PF, Olsen GN, Ganguly R, Johnson JE (1973) Immune responses of the human respiratory tract. I. Immunoglobulin levels and influenza virus vaccine antibody response. J Immunol 111:38–41

Waldman RH, Lazzell VA, Bergmann C, Khakoo R, Jacknowitz AI, Howard SA, Rose C (1981) Oral immunization against influenza. Trans Am Clin Climatol Assoc 93:133–140

Webster RG, Askonas BA (1980) Cross-protection and cross-reactive cytotoxic T cells induced by influenza virus vaccines in mice. Eur J Immunol 10:396–401

Webster RG, Laver WG (1980) Determination of the number of non-overlapping antigenic areas on Hong Kong (H3N2) influenza virus hemagglutinin with monoclonal antibodies and the selection of variants with potential epidemiological significance. Virology 104:139–148

Wells MA, Albrecht P, Daniel S, Ennis FA (1978) Host defense mechanisms against influenza virus: interaction of influenza virus with murine macrophages in vitro. Infect Immun 22:758–762

Wells MA, Daniel S, Djeu J, Kiley S, Ennis FA (1983) Recovery from a viral respiratory tract infection. IV. Specificity of protection by cytotoxic T lymphocytes. J Immunol 130:2908–2914

Welsh RM (1981) Natural cell-mediated immunity during viral infections. Curr Top Microbiol Immunol 92:83–103

White J, Helenius A, Gething MJ (1982) Haemagglutinin of influenza virus expressed from a clonal gene promoter membrane fusion. Nature 300:658–659

Wiley DC, Wilson IA, Skehel JJ (1981) Structural identification of the antibody binding sites of Hong Kong influenza haemagglutinin and their involvement in antigenic variation. Nature 289:373–378

Wilson IA, Skehel JJ, Wiley DC (1981) Structure of the haemagglutinin membrane glycoprotein of influenza virus at 3f resolution. Nature 289:366–373

Wong GHW, Clark-Lewis I, McKimm-Breschwin JL, Harris AW, Schrader JW (1983) Interferon-gamma induces enhanced expression of Ia and H-2 antigens on B lymphoid, macrophage and myeloid cell lines. J Immunol 131:788–793

Wraith DC, Askonas BA (1985) Induction of influenza A virus specific cross-reactive T cells by a nucleoprotein/haemagglutinin preparation. J Gen Virol 66:1327–1331

Wright PF, Okabe N, McKee KT, Maassab HF, Karzon DT (1982) Cold-adapted recombinant influenza A virus vaccines in seronegative young children. J Infect Dis 146:71–79

Wright PF, Murphy BR, Keruina M, Lawrence EM, Phelan MA, Karzon DT (1983) Secretory immunological response after intranasal inactivated influenza A virus vaccination: evidence for IgA memory. Infect Immun 40:1092–1095

Wyde PR, Wilson MR, Cate TR (1982) Interferon production by leucocytes infiltrating the lungs of mice during primary influenza virus infection. Infect Immun 38:1249–1255

Wylie DE, Sharman LA, Klinman NR (1982) Participation of the major histocompatibility complex in antibody recognition of viral antigens expressed on infected cells. J Exp Med 155:403–414

Yamada A, Ziese MR, Young JF, Yamada YK, Ennis FA (1985) Influenza virus hemagglutinin-specific cytotoxic T cell response induced by polypeptide produced in *Escherichia coli*. J Exp Med 162:663–674

Yap KL, Ada GL (1977) Cytotoxic T cells specific for influenza virus-infected target cells. Immunology 32:151–159

Yap KL, Ada GL (1978) The recovery of mice from influenza A virus infection: adoptive transfer of immunity with influenza virus-specific cytotoxic T lymphocytes recognising a common virion antigen. Scand J Immunol 8:413–420

Yap KL, Ada GL (1979) The effect of specific antibody on the generation of cytotoxic T lymphocytes and the recovery of mice from influenza virus infection. Scand J Immunol 10:325–332

Yap KL, Ada GL, McKenzie IFC (1978) Transfer of specific cytotoxic T lymphocytes protects mice inoculated with influenza virus. Nature 273:238–239

Yap KL, Braciale TJ, Ada GL (1979) Role of T-cell function in recovery from murine influenza infection. Cell Immunol 43:341–351

Yarchoan R, Nelson DL (1984) Specificity of in vitro anti-influenza virus antibody production by human lymphocytes: analysis of original antigenic sin by limiting dilution cultures. J Immunol 132:928–935

Yarchoan R, Murphy BR, Strober W, Clements ML, Nelson DL (1981) In vitro production of anti-influenza virus antibody after intranasal inoculation with cold-adapted influenza. J Immunol 125:1958–1963

Yewdell JW, Frank E, Gerhardt W (1981) Expression of influenza A virus internal antigens on the surface of infected P815 cells. J Immunol 126:1814–1819

Yewdell JW, Bennink JR, Smith GL, Moss B (1985) Influenza A virus nucleoprotein is a major target antigen for cross reactive anti-influenza A virus cytotoxic T lymphocytes. Proc Natl Acad Sci USA 83:1785–1789

Young JF, Desselberger U, Palese P, Ferguson B, Shatzman AR, Rosenberg M (1983) Efficient expression of influenza virus NS1 non-structural proteins in *Escherichia coli*. Proc Natl Acad Sci USA 80:6105–6109

Zahradnik JM, Kasel JA, Martin RR, Six HR, Cate TR (1983) Immune responses in serum and respiratory secretions following vaccination with a live cold-recombinant (CR35) and inactivated A/USSR/77 (H1N1) influenza virus vaccine. J Med Virol 11:227–285

Zee YC, Osebold JW, Dotson WM (1979) Antibody responses and interferon titers in the respiratory tracts of mice after aerosolised exposure to influenza virus. Infect Immun 25:202–207

Zweerink HJ, Courtneidge SA, Skehel JJ, Crumpton MJ, Askonas BA (1977) Cytotoxic T cells kill influenza virus-infected cells but do not distinguish between serologically distinct type A viruses. Nature 267:354–356

Defective Interfering Viruses and Infections of Animals*

A.D.T. BARRETT[1] and N.J. DIMMOCK[2]

1 Introduction

Defective interfering (DI) virus particles are generated during the replication of many, possibly all, animal viruses (HUANG and BALTIMORE 1977; PERRAULT 1981). They characteristically have a genome which contains deletions, often of the majority, of the standard (infectious) virus genome. These deletions mean

* The authors thank the Science and Engineering Research Council, Medical Research Council, Agricultural Research Council, The Wellcome Trust and the University of Warwick Research and Innovations Fund for financially supporting their work cited in this review

[1] Department of Microbiology, University of Surrey, Guildford, Surrey, GU2 5XH, United Kingdom
[2] Department of Biological Sciences, University of Warwick, Coventry, CV4 7AL, United Kingdom

Table 1. Properties of DI virus particles

---

1. Have a genome *generated* by deletion from the genome of standard virus
2. Use the structural proteins synthesized by standard virus; hence are antigenically identical with standard virus
3. Cannot replicate (i.e. are *defective*); hence standard virus must co-infect the same cells to *propagate* progeny DI virus
4. Reduce the yield of standard virus from co-infected cells. Thus, DI virus *interferes* with standard virus multiplication
5. During co-infection the absolute amount of DI virus and its amount relative to standard virus are *enhanced*
6. Require a functional nucleic acid for interference

---

that DI virus can only replicate in cells co-infected with standard virus. Such co-infection normally results in the enhancement of the DI virus population and a concomitant reduction in standard virus – the phenomenon of interference (HUANG and BALTIMORE 1977; HOLLAND et al. 1980; PERRAULT 1981). These and other properties of DI viruses are summarised in Table 1. The point that DI virus nucleic acid is encapsidated in the normal complement of coat proteins synthesized by standard virus should be emphasized, since it follows that DI and standard virus are antigenically identical and that they should stimulate and respond to host immune responses in the same way. The intimate biochemical dependence of DI virus on standard virus for its replication as well as encapsidation explains why, for the most part, interference is specific for the homologous standard virus. For an account of the biochemical properties of DI viruses the reader is referred to reviews by HUANG and BALTIMORE (1977), HOLLAND et al. (1980) and PERRAULT (1981).

A large amount of literature has accumulated on DI viruses, with the majority of reports (see PERRAULT 1981) concentrating on the interfering properties of DI viruses in cells in culture or on DI viruses as a tool to investigate replication. Relatively few studies have dealt with the effects of DI particles in vivo, despite the obvious extrapolations which suggest that DI viruses have the potential to act as powerful modulators of infection and disease. Some of the earliest work on DI particles was performed with incomplete virus of the A/PR/8 strain of influenza virus in ovo and in mice (BERNKOPF 1950; VON MAGNUS 1951). "Incomplete" virus had been recognised by its particle: infectivity ratio higher than that of the normal 'complete' virus. Incomplete virus was produced then, as it is today, by serial high-multiplicity passage of standard virus. Intranasal inoculation with dilutions of incomplete virus caused mice to develop a fatal pneumonia, while undiluted incomplete virus inoculated in exactly the same way resulted in fewer lung lesions, less virus production and lower mortality. Likewise, intracerebral inoculation with a high dose of incomplete virus caused fewer deaths than a 10^4-fold dilution (BERNKOPF 1950). These experiments were the first to show an effect of DI viruses on viral infections of animals. Later on, similar results were obtained by MIMS (1956) working with the bunyavirus Rift Valley fever. He observed that inoculation with incomplete virus lengthened the incubation period of the disease and reduced the extent of virus multiplication in mice. The incomplete virus was also found to be capable of immunising mice, since high titres of neutralizing antibody were obtained after its administration.

The idea that DI particles could play a role in the expression of naturally occurring viral diseases was postulated by HUANG and BALTIMORE (1970), who proposed that DI viruses reduced the virulence of acute infections and helped to establish and maintain persistent infections. The early studies in vitro on which this extrapolation was based have been amply confirmed (for a review see HOLLAND et al. 1980), but the relevance of this body of work to the more complex in vivo situation is still unclear.

This review aims to draw together data on the effects of DI viruses on infections in vivo in order to bring out general principles and pointers for future directions of research. At the same time we shall examine the underlying assumption that DI viruses protect animals in the same way that they protect cells in culture.

2 Effect of DI Virus on Disease Processes

Experiments by DOYLE and HOLLAND (1973) and HOLLAND and DOYLE (1973) first showed that purified DI virus (vesicular stomatitis virus: VSV) could prevent the lethal encephalitis in young adult mice caused by intracerebral inoculation of standard virus. They found that large quantities of DI particles (5×10^{10}) were required to prevent death (i.e. to give protection), and that this was sufficient to protect against only low doses of standard virus. With higher doses of standard virus, DI VSV only delayed the time of death, but the most important finding was that the otherwise rapidly fatal disease (death 2–3 days post inoculation) was converted to a more slowly progressive infection with wasting, hind limb paralysis and death after 7–9 days post inoculation. DOYLE and HOLLAND (1973) showed that DI virus reduced standard virus multiplication in the brain, but they failed to directly detect DI particles in such brains. A further study using newborn mice by HOLLAND and VILLARREAL (1975) showed that DI VSV did multiply in the brains of these younger animals, although only to low titres, and it was necessary to use pooled material in order to detect DI particles. Identical results were obtained with DI rabies virus in newborn mouse brains. Since DI VSV particles are shorter than standard virus, it was possible to characterise those produced in the mouse brains, and to show that they were identical in length to those in the inoculum and therefore likely to have been propagated from the inoculum rather than generated de novo. It appears that the brains of newborn mice propagate DI VSV like a tissue culture system, but that adult brains behave very differently. One possible explanation is that increased myelination in adults may have impeded neurone-to-neurone spread of the virus. HOLLAND's group concluded that the survival of mice after an otherwise lethal dose of VSV was due to interference mediated by DI virus, and distinguished this effect from the presumably immunological protection conferred by the administration of DI particles 10 days prior to standard virus challenge.

RABINOWITZ et al. (1977) used VSV to confirm and extend the studies of Holland and co-workers. Following the intracerebral co-inoculation of 3- to 4-week-old mice with DI and standard VSV, they observed a slowly progressive disease of the central nervous system which differed in a number of pathological aspects from that induced with standard VSV alone. Mice still died, but survived 5–9 days post inoculation, compared with 2–3 days with standard VSV alone. Although in their earlier work HOLLAND and colleagues did not study host immune responses in DI VSV-treated mice or use inactivated preparations to control for possible immunogenic effects of DI virus, JONES and HOLLAND (1980) and FULTZ et al. (1982a) demonstrated that biologically active DI particles, but not UV-inactivated DI or UV-inactivated standard virus, protected mice against challenge by standard VSV. Hence protection was not due to an immune response to virus coat proteins, and their earlier conclusions were vindicated; the cause of the heterologous interference found by CRICK and BROWN (1977) remains unclear, but may be explained by the particular preparation of DI virus used. Some in vitro studies with VSV and DI VSV in cultured neurones also support the results of HOLLAND's group. FAULKNER et al. (1979) reported that co-infection of mouse neurones in culture delayed their death and suppressed virus growth, suggesting that DI particles may exert an effect on viral infections of the central nervous system by interfering with the multiplication of standard virus. In addition, protection in mice has also been demonstrated with rabies virus, another rhabdovirus (KOPROWSKI 1954; WIKTOR et al. 1977).

Although many of the early studies used VSV, modulation of infection in vivo has also been demonstrated with other virus systems. Inoculation with the DI WSN together with A/WSN strain (H1N1) of influenza virus by the intracerebral route showed that adult mice were protected from lethal infection and that virus yields were diminished (GAMBOA et al. 1976), thus confirming earlier studies by BERNKOPF (1950). However, HOLLAND and DOYLE (1973) failed to demonstrate protection. Unfortunately, GAMBOA et al. (1976) did not include controls for the immunogenic effects of the inoculum, and protection may have resulted from stimulation of the immune response. Prevention by DI virus of fatal pneumonia in mice following intranasal inoculation of the A/PR/8 (H1N1) strain of influenza virus has been reported by BERNKOPF (1950), VON MAGNUS (1951) and RABINOWITZ and HUPRIKAR (1979). DI virus-treated animals developed a typical but mild form of the disease which was self-limiting, and virus yields were diminished. HOLLAND and DOYLE (1973) confirmed the reduction in virus yield, but failed to demonstrate protection.

While most reports concluded that DI virus interfering activity was responsible for the amelioration, RABINOWITZ and HUPRIKAR (1979) were of the opinion that protection was caused by augmented humoral immune responses. The lack of stringent control of the amounts of immunogen inoculated makes it impossible to draw firm conclusions. However, in recent studies using the same mouse model DIMMOCK et al. (1986) have demonstrated that the lethal pulmonary infection of mice by A/WSN can be prevented in 80% of mice by active DI virus. DI virus treated with beta-propiolactone, which destroys interfering activity as measured in vitro but leaves the haemagglutinin (HA) and neuraminidase activities unaffected, was the control, and this did not alter the course of infection. However, the surprising result of these experiments was that although DI virus prevented death it did not affect the multiplication of virus in the lungs, as shown by assay of infectivity, total (HA) antigen and neuraminidase activity, or distribution of viral antigens in lung tissue. Such mice became ill and developed the same type of cellular infiltration in the lung as those with the lethal infection, but consolidation was delayed, less extensive and resolved in due course. Since the pathology in the mouse is immune (T cell) mediated the authors suggested that as there was no affect on virus multiplication DI virus appeared to be inhibiting the deleterious T cell response (see Sect. 6.4). This unique situation, the first instance of a DI virus "interfering" without affecting virus multiplication, remains to be clarified. Interferons were not obviously involved, as local lung levels were not affected by DI virus.

WELSH et al. (1977) have shown that intracerebral administration of DI lymphocytic choriomeningitis virus (LCMV) prevented central nervous system disease and death caused by standard virus in 2-day-old rats. The synthesis of standard virus and LCMV antigens was reduced, but no interferon or host immune responses appear to have been involved in protection. Since the disease caused by standard virus is immune-mediated, it was suggested that protection correlates with the ability of DI LCMV to inhibit the expression of LCMV surface antigens in infected cells (WELSH and OLDSTONE 1977). HELP and COTO (1980) have shown that DI virus is generated in newborn mice infected with another arenavirus, Junin, and propose that the DI virus is responsible for the delay in death and reduced yield of standard virus in co-infected animals.

The generation, replication and in vivo interfering ability of DI reovirus have been demonstrated in newborn rats after intracerebral or subcutaneous inoculation (SPANDIDOS and GRAHAM 1976). Surviving rats were runted and chronically infected with the brain and liver of most animals yielding virus up to 60 days post inoculation. However, none was found at 80 days post inoculation. Apart from two mutants these viruses had the same temperature-sensitive phenotype as the inoculum. In the initial stages of infection DI particles were said to lack only the L_1 RNA segment, but later isolates had more segments deleted. The claim that DI virus was generated de novo is difficult to substantiate, as it cannot be proved that the inoculum was free of DI virus, but it is clear from changes in the genome that DI virus was evolving during the persistent infection. SPANDIDOS and GRAHAM also report that DI virus could be isolated when infectious virus was no longer present, and suggest that it was propagated by complementation between DI genomes carrying different deletions, an interesting suggestion which has yet to be validated experimentally.

Only in the Semliki Forest virus (SFV) system have controls to monitor for the effects of possible immune responses to the additional virus antigen represented by DI virus been consistently used. DIMMOCK and KENNEDY (1978) observed that co-inoculation with DI and standard SFV by the intranasal route resulted in protection of mice from the lethal encephalitis caused by virulent standard SFV. Administration of an amount of UV-inactivated standard virus equivalent in HA units to that of DI virus had no effect on the progress or outcome of infection. Mice given DI virus either survived infection without showing any clinical signs of standard virus infection or died following the normal pattern of SFV disease. Simultaneous inoculation of the DI virus and the standard virus was necessary for optimum protection. Inoculation with DI virus 2 h before infection also gave protection, but inoculation 2 h after infection had no effect. DI SFV reduced the multiplication of standard virus in the brain by at least 10^5-fold, and DI virus was detected in the brain after amplification in tissue culture. Protection was not due to the induction of interferon or stimulation of the host immune responses.

Later studies also attested to the specificity of protection, as DI SFV did not protect mice against infection by heterologous viruses (encephalomyocarditis, West Nile or Japanese encephalitis viruses, or the L10 strain of SFV; see Table 2). Table 2 also shows that even when heterologously infected mice were treated with DI SFV *and* SFV in order to propagate the DI virus, there was no heterologous protection. Thus, protection did not appear to involve the induction of non-specific defence responses (macrophages, natural killer cells, interferons, etc.; see BARRETT and DIMMOCK 1984b, unpublished data). All the evidence suggested that the protection of the mice was due to the interfering capacity of the DI virus. Further studies by CROUCH et al. (1982) and BARRETT et al. (1984b) have shown that the brains of protected mice have no pathological or histochemical lesions whatsoever, nor any evidence of immune cell infiltration. Again, these results would seem to support the view of DIMMOCK and KENNEDY (1978) that DI SFV protects mice by interference rather than by stimulation of the host's immune processes by DI virus proteins.

In comparison to the many studies of infections of the central nervous system, FULTZ et al. (1982a) have reported on VSV infection of Syrian hamsters after intraperioneal inoculation. This route results in pathology of the lymphoreticular system (with major necrosis of the splenic periarteriolar lymphoid sheath), but little or no involvement of the central nervous system. Nonetheless, inoculation with biologically active DI VSV protected hamsters against the lethal standard virus infection, but in contrast to studies on inoculation of the central nervous system of mice, much smaller numbers of DI particles were required to achieve a significant effect (see Sect. 3). Protection resulted in low levels of standard virus in the serum and tissues. FULTZ et al. (1982a) suggest that protection is not only mediated by the interfering capacity of DI virus but that other factors such as interferon are involved, since they demonstrated heterologous interference by DI virus of the Indiana serotype with the New Jersey serotype of VSV and found that interferon was induced by DI VSV in the mice. Also reported is the unique finding of protection when DI and standard virus were inoculated by different routes.

Finally, in this last paragraph we wish to re-examine the question of how DI virus modulates disease in vivo. The in vitro situation is understandable, because the situation can be arranged so that the majority of cells are co-infected by standard and DI virus and interference is maximized. The situation in vivo is not so clear, since with the very large number of cells present in an animal the chances of co-infection seem remote, especially following intracerebral and intraperitoneal inoculation. It is possible that some "funnelling" takes place, whereby the virus is transported to the particular cells for which standard and

Table 2. Failure of DI SFV (prepared from *SFV* ts[+]) to exert heterologous protection in vivo[a]

Inoculum[b]	Mice surviving (%)	Mean day of death
SFV[c]	0	4.9
SFV + UV SFV	0	5.1
SFV + DI SFV	50	5.1
L1O	0	4.0
L1O + UV SFV	0	4.0
L1O + DI SFV	10	4.6
EMC[d]	0	4.1
EMC + UV SFV	0	4.1
EMC + UV SFV + SFV	0	4.9
EMC + DI SFV	0	4.6
EMC + DI SFV + SFV	0	5.1
JEV	0	8.5
JEV + UV SFV	0	8.5
JEV + UV SFV + SFV	0	5.2
JEV + DI SFV	0	8.6
JEV + DI SFV + SFV	0	6.7[e]
WNV	0	6.2
WNV + UV SFV	0	6.4
WNV + UV SFV + SFV	0	5.4
WNV + DI SFV	0	6.6
WNV + DI SFV + SFV	0	5.9

[a] Ten mice were inoculated intranasally with DI SFV at -2 h and together with 10 LD_{50} at time zero of the viruses shown in the Table (DIMMOCK and KENNEDY 1978)

[b] Viruses are: SFV, Semliki Forest virus (ts[+] strain); L1O strain of SFV; WNV, West Nile virus; JEV, Japanese encephalitis virus; EMC, encephalomyocarditis virus; UV SFV, non-infectious, ultraviolet-irradiated standard SFV diluted to the same number of HA units as DI SFV; DI SFV, DI virus of the ts[+] strain

[c] pfu/LD_{50} for SFV ts[+] = 600; for SFV L10 = 126; for EMC = 100; for JEV = 2000; and for WNV = 20

[d] EMC virus data from BARRETT and DIMMOCK (1984b)

[e] Average value conceals bimodal distribution with deaths from SFV at 5 days and from JEV at 8 days

DI virus are tropic, and it is in these cells that interference would take place (FULTZ et al. 1982a). Alternatively, the "funnelling" in the case of intranasal inoculation and resulting infection of the central nervous system may have a physical basis, since the ratio of olfactory nerve fibres to the underlying layer of mitral and tufted cells is 300:1 (ALLISON and WARWICK 1949). Other mechanisms may be at work, as it seems that, in addition to interference at the molecular level, DI viruses influence pathogenesis via interference with the immune responses, and this argument will be presented in detail in Sect. 6.

Summary. There is now overwhelming evidence from model experimental systems that DI viruses can modulate disease processes in a variety of ways. Discussion of how they achieve this and whether or not DI viruses play a role in natural disease will follow in the remainder of this review.

3 Properties of DI Viruses In Vivo

The amounts of both standard virus and of DI virus inoculated are important in determining the extent of protection. In general, the larger the quantity of DI virus administered, the greater the effects observed; this presumably reflects the greater chance of cells being co-infected by DI and standard virus, and therefore of interference. Against the run of evidence, CAVE et al. (1984) report that there was an optimum concentration of DI VSV and that protection was diminished when larger amounts were used; in a later paper they also observed that the extent of protection did not correlate with the ratio of DI to standard virus or with the total amount of DI virus inoculated (CAVE et al. 1985). The use of old mice past breeding age (9–10 months) may possibly have some bearing on these unexpected results.

HOLLAND and co-workers have shown the large quantities (5×10^{10} particles) of DI virus were required, and that when these were inoculated by the intracerebral route they protected against small (250 pfu), but nonetheless fatal, doses of standard virus (DOYLE and HOLLAND 1973). Co-inoculation with higher doses of standard virus resulted in an increase in survival time only, and death ensued. Similarly, RABINOWITZ et al. (1977) used 10^{11} DI particles to achieve modulation of infection by 1.5×10^5 pfu standard virus. A minimum of 3×10^8 particles of active DI virus given by the intracerebral route are required to protect adult mice (HOLLAND et al. 1978), but 10^3 DI particles given intraperitoneally protected hamsters (FULTZ et al. 1982a). The SFV system has shown that protection is proportional to the ratio of DI and standard virus in the inoculum (DIMMOCK and KENNEDY 1978; BARRETT and DIMMOCK 1984c). Here, an amount of DI virus equivalent to 2×10^5 pfu (roughly 10^7 DI particles) was required to protect the majority of mice against 10 LD_{50} (6×10^3 pfu) standard virus.

One surprising finding has been that the type of DI virus used in experiments can cause completely different manifestations of disease and protection. JONES and HOLLAND (1980) reported a quantitative difference in the efficacy of DI VSV viruses, in that ten times more DI virus "MS" was required than of DI virus "CAR 4" to achieve the same degree of protection against standard VSV inoculated intracerebrally. Similarly, FULTZ et al. (1981) reported that DI virus LT protects mice to greater degrees than does DI virus 2 against a systemic VSV infection. Since these different isolates of DI VSV have different RNA sequences and structures (e.g. DI-CAR 4 is completely self-complementary, while DI-MS is only partially self-complementary), it appears that the sequence content of DI virus nucleic acid is important for protection. However, the biological properties of DI VSV isolates are not infinitely variable at this level, since those derived independently by DOYLE and HOLLAND (1973) and RABINO-

Table 3. Comparison of some properties of DI SFV preparations.
(From BARRETT and DIMMOCK 1984b)

	p5	p4	p13
Interference in vitro	+	+	+
Protection in vivo	−	+	+
Immunity to SFV challenge	NA	−	+

p refers to the number of undiluted passages in BHK cells

WITZ et al. (1977) produced the same changes in standard VSV infection (delay in death and hind limb paralysis).

Other studies have shown more clearly that DI viruses differ qualitatively in their biological properties. While some isolates of DI SFV protected mice, others which had the same interference titre in vitro did not (BARRETT and DIMMOCK 1984b). Furthermore, DI SFV preparations which protected identical numbers of mice differed in the effects they had upon the mouse. Treatment with DI SFV p13 (produced by 13 serial undiluted passages in BHK cells) left surviving mice immune to a subsequent lethal challenge with SFV; this was to be expected, as the virus multiplied systemically. However, infected mice treated with DI SFV p4 (produced by four serial undiluted passages) were all susceptible to SFV challenge at 3 weeks post infection (Table 3). Thus, DI SFV p4 induced no protective immune response after infection, even though for the first 2 days the virus multiplied to normal levels throughout the animal. Differences in DI SFV preparations were not random. "Sister" DI SFV preparations at the same passage level and from the same parental inoculum (e.g. DI SFV p13a, p13b, etc.) all behaved similarly, while changes in passage levels (e.g. to p14) affected the biological activity of the DI virus. Surprisingly, a single passage radically altered the biological properties of a DI virus preparation from "protecting" to "non-protecting" or vice versa (BARRETT and DIMMOCK 1984b). Since the quantity of DI virus in each preparation was very similar, it was concluded that DI viruses were qualitatively different and that only some have the ability to offer protection in vivo. Data were consistent with the hypothesis that each DI virus passage level contained a mixture of several genetically different DI viruses and that at each new passage a different DI virus became predominant (KÄÄRIÄINEN et al. 1981; BARRETT et al. 1984a).

Summary. There are wide variations in the in vivo properties of different DI virus isolates which are in accord with the known diversity of DI genomic sequences and structures as exemplified by VSV, SFV, influenza and Sindbis viruses (PERRAULT 1981; NAYAK and SIVASUBRAMANIAN 1983; O'HARA et al. 1984; LEHTOVAARA et al. 1981; MONROE et al. 1982; FIELDS and WINTER 1982; JENNINGS et al. 1983) and their biological activities in vitro (BARRETT et al. 1984a; BARRETT and DIMMOCK 1984a). It also seems likely that only some DI RNAs have a sequence which enables them to protect animals in vivo. Clearly, the relationship between structure and function of DI virus genomes is an area which needs to be investigated.

4 Properties of the Host

Rodents have most commonly been used to study the modulatory effects of DI virus, and the majority of studies have been performed in mice.

4.1 Age and Protection

The age of the animal is critical in determining the efficacy of protection, since DI virus is more effective in adult than in newborn or weanling mice. GAMBOA et al. (1976) found that intracerebral inoculation of DI and standard influenza virus (A/WSN) into 7-week-old Swiss albino mice resulted in protection, while in 3-week-old mice virus titres were only reduced and death ensued. A similar effect of age was found also with intranasal inoculation (RABINOWITZ and HUP-RIKAR 1979). While DI VSV completely protected adult mice from 10^2 pfu inoculated intracerebrally, only 12% of neonates so treated survived, the remainder showing delayed onset of death similar to that seen with a higher standard: DI virus ratio in adults (HOLLAND and VILLARREAL 1975). This was surprising in view of the prolific generation of DI particles in neonates (Sect. 4.2). In the SFV system the extent of protection varied between strains of mice, and this appeared to reflect in part the pfu; LD_{50} ratio of standard virus. However, the random-bred strains of mice with the same pfu:LD_{50} ratio differed in the extent of protection even when identical quantities of DI virus were inoculated (BARRETT and DIMMOCK 1984c).

4.2 Age and Generation of DI Virus

The generation of DI virus in animals is also variable, although as it is difficult to be sure that the inoculum contains *no* DI virus particles, generation and propagation cannot be distinguished. Because DI and standard rhabdovirus particles can be separated so easily, this provides by far the best example. In both the VSV and rabies virus systems, DI particles are extensively generated in baby mice and poorly in adults (HOLLAND and VILLARREAL 1975). Inoculated DI LCMV, however, was propagated in both newborn and adult mice (POPESCU and LEHMANN-GRUBE 1977). Subcutaneous injection of 2-day-old rats resulted in an encephalitis from which up to 10% died. About half the survivors cleared the virus and grew normally, but the others were chronically infected and runted. Defective virus (not shown to be interfering) was isolated from the latter after an amplification step at intervals up to 100 days post inoculation, at which time infectious virus could no longer be found. For the record, DI particles of the arenavirus Junin were generated/propagated in newborn mice (HELP and COTO 1980) and DI equine herpesvirus type 1 in adult hamsters by serial propagation in vivo (CAMPBELL et al. 1976).

4.3 Virulence, DI Virus and the Genetic Constitution of the Host

One of the original suggestions of HUANG and BALTIMORE (1970) was that a factor in the expression of virulence might be the generation and subsequent action of DI virus, and there is indeed evidence that the virulence of flaviviruses for certain strains of mice is determined by the presence of DI virus. DARNELL and KOPROWSKI (1974) proposed that the ability to generate DI particles was responsible for the genetic resistance of mice to West Nile virus infection initiated by the intraperitoneal route, since cell cultures from resistant strains of mice produced interfering particles whereas cells from susceptible mice did not. BRINTON (1983) confirmed these studies and showed that cell cultures from resistant mice generate and amplify DI particles more efficiently than cell cultures from susceptible mice. A similar conclusion was reached by SMITH (1981), who implicated "interfering virus" (not defined as DI) as being involved in the genetic resistance of mice to infection of the central nervous system by Banzi, another flavivirus. SMITH (1981) found that after intraperitoneal inoculation mice resistant to infection had high levels of interfering virus, while susceptible strains of mice had much lower levels. Surprisingly, interfering virus was detected in the spleen and not the brain, and the amount of interfering virus was increased by the presence of cyclophosphamide. On the basis of these results, SMITH (1981) proposed that interfering virus originated in cells of the lymphoreticular system. Virulence of rabies viruses could not be correlated with their ability to generate DI virus particles (WUNNER and CLARK 1980), despite the fact that rabies virus infections are susceptible to modulation by DI virus (KOPROWSKI 1954; WIKTOR et al. 1977).

Summary. The virulence of a virus (or its converse, the genetic resistance of a host) is, at least in part, dependent on the generation of DI virus particles. This seems as much a host-controlled property as a virus-controlled one. However, evidence at present is limited to flavivirus systems, and the virulence of rabies virus cannot be explained in these terms.

4.4 Cell Susceptibility and DI Virus

In vitro experiments show clearly that the ability to generate DI viruses depends on the particular cell line infected. Mouse 3T3, rat NRK, chick embryo fibroblast and baby hamster kidney cells generated DI SFV readily, while HeLa cells failed completely even after 200 undiluted passages (STARK and KENNEDY 1978). Similarly, the ability of DI SFV to interfere varied quantitatively with the type of cell in which the assay was conducted (BARRETT et al. 1981). HeLa cells did not generate DI VSV de novo or propagate added DI VSV, but interference with standard VSV multiplication did take place (HOLLAND et al. 1976a). Thus, it is conceivable that in vivo the generation and interfering effects of DI virus will depend upon the particular type of differentiated cell(s) in which the virus multiplies or to which DI virus gains access.

5   The Route of Administration of DI Virus

The majority of studies have involved infection of the central nervous system resulting from either intracerebral or intranasal inoculation. Many have used the intracerebral route, since only small numbers of standard virus particles are required to kill the animal and this optimizes the prophylactic effects of DI particles. Unfortunately, this route is most unnatural and causes trauma for the animal, including disruption of the blood–brain barrier, so that 90% of the inoculum is distributed systemically (MIMS 1964). Despite these drawbacks, protection experiments using this route have been very effective (DOYLE and HOLLAND 1973; HOLLAND and DOYLE 1973; GAMBOA et al. 1976; RABINOWITZ et al. 1977; SPANDIDOS and GRAHAM 1976; WIKTOR et al. 1977; WELSH et al. 1977; JONES and HOLLAND 1980) but not with SFV in mice (N.J. DIMMOCK, unpuplished data).

Other workers have used the intranasal route to initiate infections of the central nervous system, as it is more natural, involves less trauma and can result in the transport of the virus along cranial nerves into the brain. The olfactory nerve endings in the olfactory mucosa are extensions of the brain itself and make unbroken (synapse-free) communication between the outside environment and the central nervous system. A series of studies with DI SFV using this route, performed by Dimmock and co-workers, have demonstrated the complete abrogation of clinical disease resulting from an otherwise lethal dose of SFV (DIMMOCK and KENNEDY 1978; CROUCH et al. 1982; BARRETT and DIMMOCK 1984b; BARRETT et al. 1984b). Early studies by HOLLAND and DOYLE (1973) with VSV using the intranasal route of inoculation failed to modulate infection, but CAVE et al. (1984, 1985) have recently demonstrated protection.

There have been only three reports of DI viruses modulating infection when inoculated systemically. SPANDIDOS and GRAHAM (1976) prevented a lethal reovirus infection in 2-day-old rats inoculated subcutaneously, while FULTZ et al. (1982a) showed that Syrian hamsters could be protected against the lethal lymphoreticular disease resulting from intraperitoneal inoculation with VSV by DI virus also inoculated intraperitoneally. Modulation has also been reported following intraperitoneal inoculation with DI SFV (BARRETT and DIMMOCK 1984d); death was delayed, but the majority eventually succumbed. Administration of DI virus at one anatomical site (intraperitoneal) and standard virus at another (intranasal) did not modulate infection with SFV (A.D.T. BARRETT and N.J. DIMMOCK, unpublished observations). This tends to support the view that DI virus must co-infect a cell with standard virus for protection to take place, as does the enhanced protection observed when DI and standard LCMV were clumped together before inoculation (WELSH et al. 1977). Protection by DI VSV given intravenously against VSV inoculated intraperitoneally (FULTZ et al. 1982a) is at variance with this notion, and may reflect the funnelling of viruses to the same target cell referred to earlier (Sect. 2). Intranasal inoculation has also been used to study respiratory infections by influenza virus and the effects of DI virus on the disease process. The early work of BERNKOPF (1950) and VON MAGNUS (1951), which was followed by the later studies of

HOLLAND and DOYLE (1973), RABINOWITZ and HUPRIKAR (1979) and DIMMOCK et al. (1986), has already been described in Sect. 2.

Summary. DI virus has successfully modulated infections when co-inoculated with standard virus by intracerebral, intranasal or intraperitoneal routes. There is one report of DI virus-mediated protection when standard virus was inoculated at a different site.

6 Modulation of the Immune Response by DI Virus

6.1 Biologically Active DI Genomes, not Immunogen, Are Mediators of Protection

Although there have been a number of studies on DI viruses in animals not many have included controls to monitor for the possible immunogenic effects of DI virus, which was always inoculated in large amounts, and the problem was highlighted by CRICK and BROWN (1977), who cast doubt on the claim that the mechanism of the protection was DI VSV-mediated interference. They found that mice could be protected by inoculation of DI VSV which had been treated with acetyl ethyleneimine, an agent which inactivates nucleic acid function, and, furthermore, that DI VSV protected against heterologous strains of VSV, rabies and a neurotropic strain of foot-and-mouth disease virus. CRICK and BROWN concluded that the protection observed was due to activation of the host defence responses and not to interference by DI virus. Despite these conclusions, a number of carefully controlled studies have demonstrated that DI viruses can protect in vivo in a manner which requires the integrity of the DI virus genome and is independent of the amount of antigen associated with the DI virus: JONES and HOLLAND (1980) and FULTZ et al. (1982a) with VSV; DIMMOCK and co-workers (DIMMOCK and KENNEDY 1978; BARRETT and DIM-MOCK 1984a, b, c, d) with SFV; and DIMMOCK et al. (1986) with influenza virus.

6.2 Not All DI Virus Isolates Protect In Vivo

The idea that not all DI virus isolates protect in vivo largely stems from data which demonstrate that there are DI viruses which do *not* modulate infection in animals, and extends the findings of JONES and HOLLAND (1980) and FULTZ et al. (1981) that on a particle basis VSV DI-LT was about 100-fold more effective in protecting hamsters than was DI-2. To cite one example, BARRETT and DIMMOCK (1984b) showed that while DI SFV isolates p4 and p13 completely abrogated disease and death from a lethal dose of SFV, DI SFV p5 did not alter the course of disease, even though they all had similar in vitro interference titres and comprised similar amounts of virus antigen (Table 3). Clearly, although p5 had interference activity in vitro, it lacked the type of biological

Table 4. Interference in vitro by DI SFV (ts$^+$ strain) with the L10 strain of SFV without cross-protection in vivo

Virus	Interference		
	In vivo[a]	In vitro	
	Survivors	RSIA titre (DIP/ml)	YRA titre (DIP/ml)
ts$^+$	5/10	$10^{5.9}$	$<10^{5.1}$
L10	1/10	$10^{5.7}$	$<10^{5.1}$

RSIA, RNA synthesis inhibition assay (BARRETT et al. 1981); YRA, yield reduction assay (BARRETT et al. 1984b); DIP, defective interfering particles
[a] 10 LD$_{50}$ of ts$^+$ is 6000 pfu and of L10 is 1260 pfu. See Table 3 for protocol

activity necessary to prevent disease. Further studies showed that DI SFV of the ts$^+$ strain of SFV would protect mice from infection by standard virus of the ts$^+$ strain, but not of the L10 strain, of SFV. However, the same DI virus preparation interfered with the replication of both ts$^+$ and L10 strains of standard virus in vitro (Table 4 and A.D.T. BARRETT, T. ATKINSON, A.R. GUEST and N.J. DIMMOCK, unpublished data). Clearly, protection of mice is a more specific process than interference in vitro.

What could be the nature of this activity which seems irrelevant to in vitro interference? One possibility is that the interfering activity, which is known to be in part dependent upon the cell (BARRETT et al. 1981), is not exerted in the tissues which are infected in vivo. Another possibility is by stimulation of the host defence system (not through the immunogenic properties of the DI virus itself, as these are identical with those of standard virus), in the sense that the interaction of an active DI genome with elements of the immune system may lead to an improved ability of the host to cope with the developing infection. How it does so remains to be determined.

DI SFV p4 provides evidence that adaptive immune responses are not stimulated to convert the normal SFV encephalitis into a subclinical infection (BARRETT and DIMMOCK 1984b), since survivors had no immunity against a subsequent lethal challenge by standard SFV. (By comparison, mice surviving through administration of another DI SFV isolate, p13, did develop a solid immunity – see Table 3.) To explain these observations it was suggested that, in addition to its intrinsic interfering activity, DI SFV p4 also had the ability to stimulate elements of non-adaptive immunity, such as macrophages, natural killer cells and interferons. However, DI SFVs do not appear to be efficient stimulators of interferon, and high levels of interferon were seen only in the brains of mice where multiplication of standard virus was unaffected by DI SFV (Fig. 1 and A.D.T. BARRETT, A.R. GUEST, T. ATKINSON and N.J. DIMMOCK, unpublished data).

In other systems, the induction of interferon may well be an important aspect of the modulating activity of DI virus. FULTZ et al. (1982a) proposed

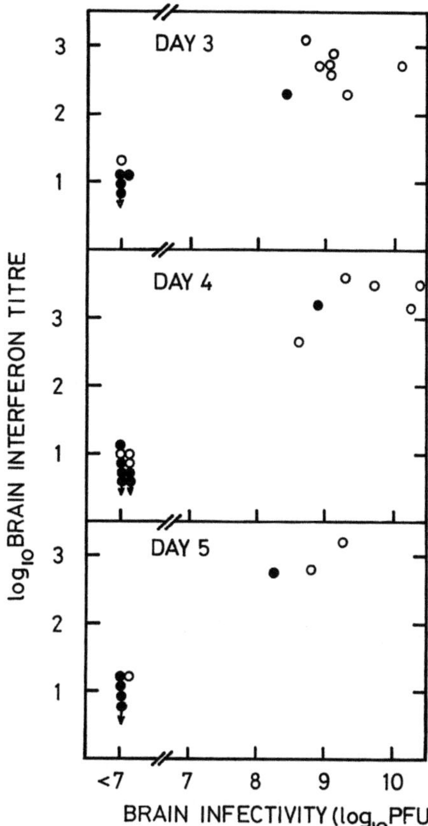

Fig. 1. Mice protected from a potentially lethal SFV infection by treatment with DI SFV p13 do not have high brain interferon titres. •, mice infected intranasally with 10 LD$_{50}$ SFV (ts$^+$)+ DI SFV; o mice infected intranasally with 10 LD$_{50}$ SFV (ts$^+$)+UV SFV. UV SFV and DI SFV contained the same concentration of virus antigen. The non-protected mice died on day 5; the others showed no clinical signs at any time. DI SFV had been prepared by 13 undiluted passages in BHK-21 cells. For details of experiment see BARRETT and DIMMOCK (1984b) and BARRETT et al. (1984b)

that DI VSV induces interferon and that it is this which mediates protection, although whether the interferon is acting as an immune modulator or is directly antiviral is unknown. They demonstrated that DI VSV (Indiana strain) interfered heterologously in hamsters with the New Jersey serotype of VSV, possibly as a result of the interferon which is induced. Similarly, in the VSV hamster model protection and persistence could also be achieved by administration of poly I:poly C (FULTZ et al. 1982a, b). In contrast, others have failed to demonstrate the induction of interferon in protected animals (see Fig. 1; DOYLE and HOLLAND 1973; WELSH et al. 1977).

6.3 Inhibition of Antigen Expression at the Cell Surface

DI viruses can also have beneficial effects where a disease frustrates the immune system. If the DI virus results in reduced synthesis of standard virus structural proteins and a concomitant reduction in the quantity of virus antigens on the surface of infected cells (WELSH and OLDSTONE 1977; HUANG et al. 1978), this may result in reduced susceptibility of a virus infection to immune attack by antibody plus complement or cytotoxic T lymphocytes, as has recently been suggested by in vivo studies with DI LCMV (BIRON et al. 1984). Recently, studies on the co-infection of BHK-21 cells with Sendai and DI Sendai virus have shown reduced expression of the HN glycoprotein at the cell surface. This was the result of faster turnover of the glycoprotein at the cell surface caused by DI particles (ROUX and WALDVOGEL 1983; ROUX et al. 1984, 1985). Such a mechanism would assist a virus infection to avoid immune surveillance.

6.4 Immunosuppression by DI Viruses

Work with different isolates of DI SFV has shown that one of these (p4) is immunosuppressive, since no protective immune response has been detected even though infectious virus is present in normal amounts up to 2 days post infection in the serum, spleen and central nervous system, and in reduced amounts for several days thereafter (BARRETT and DIMMOCK 1984b; BARRETT et al. 1984b). There is no information about how such a phenomenon is effected, but it is easy to imagine how immunosuppression could go hand in hand with the establishment by DI viruses of persistent infections (FULTZ et al. 1982b, 1984). Indeed, a proportion of mice surviving a lethal dose of SFV by administration of DI SFV p4 did establish a persistent infection from which infectious virus could be directly isolated (ATKINSON et al. 1986). More mice became persistently infected with the "immunosuppressive" DI SFV p4 than with the "non-immunosuppressive" DI SFV p13. It is also interesting to note that the persistence of infectious pancreatic necrosis virus (IPNV) in salmon is immunosuppressive, and that IPNV is frequently observed to produce DI particles in tissue culture (MACDONALD and YAMAMOTO 1978).

In this discussion of immunosuppression by DI viruses it is useful to draw on the analogy with defective retroviruses. ECKNER and HETTRICK (1977) have already suggested that the spleen focus-forming virus (SFFV) is a DI virus, pointing out that it is not only defective, requiring the helper Friend lymphatic leukaemia virus (LLV-F) for replication, but that it also interferes with XC plaque production by LLV-F in several different strains of mouse embryo cells (ECKNER and HETTRICK 1979). In chickens, the replication-defective reticuloendotheliosis virus (REV) strain T, together with its helper virus, REV-A, induces a specific suppressor cell population which acts on cytotoxic cells capable of lysing REV-T tumour cells (RUP et al. 1979; WALKER et al. 1983). This is at least partly due to the activity of the defective REV-T, since REV-T-infected cells alone induce or activate a suppressor cell population. However, cells containing REV-T genomes neither synthesize virus polypeptides nor release virus particles (HOELZER et al. 1980); hence it is not known how they interact with the immune system. At the moment, no more than a parallel can be drawn between DI viruses and defective retroviruses, but the possibility that both are capable of immunomodulation makes an interesting comparison.

A more clear-cut example of DI virus-mediated immunosuppression is inferred from the protection of C_3H mice against a lethal influenza pneumonia (DIMMOCK et al. 1986). The pathology is immune-mediated, as shown by the delay in death which ensues when infected mice are immunocompromised (HURD and HEATH 1975; SUZUKI et al. 1974; SULLIVAN et al. 1976; WYDE et al. 1977) and the increase in mortality when mice receive primed T (delayed hypersensitivity: T_d) cells (LEUNG and ADA 1980; ADA et al. 1981; LIEW and RUSSELL 1983). Thus, in protecting mice the DI virus could either be inhibiting virus multiplication or suppressing the deleterious T_d cell immune response. In fact, the kinetics of virus multiplication and virus glycoprotein antigen production and location are unaffected by DI virus; it thus follows that the DI virus suppresses host immunity. Suppression appeared to be quantitative, as evidenced by the decrease in the size of the pathological lesions but not their constitution. That an active DI virus genome was required was shown by the failure of beta-propiolactone-treated DI virus to protect mice.

A second change in immune function in DI WSN virus-treated mice was a large (tenfold) increase in local lung haemagglutination-inhibiting (HI), nonneutralizing antibody. Since there was no difference in lung HA antigen between the two experimental groups, it appears that DI influenza stimulates the B cell response. However, whether there is stimulation of T helper cell function or inhibition of T suppressor cell formation has yet to be determined, as has the possibility that the HI antibody prevented T_d cells from recognising cell surface HA.

Summary. Not all DI viruses are able to protect animals in vivo, even though they apparently have strong interfering activity in vitro, and this may reflect a tissue- or cell-specific inability of these DI viruses to express their interference. However, there is evidence that DI viruses that protect do so because they are able to modulate elements of the host's immune system. This is envisaged as a property additional to molecular intracellular interference, and which requires the participation of an active DI genome, not DI virus, as an immunogen. Such a hypothesis is preliminary, and more evidence is needed to support or refute it.

7 Distribution of Standard and DI Virus in DI Virus-Protected Animals

7.1 Standard Virus

Infectivity titres in tissues of protected mice are either undetectable (DOYLE and HOLLAND 1973; SPANDIDOS and GRAHAM 1976; BARRETT et al. 1984b); reduced (HOLLAND and DOYLE 1973; SPANDIDOS and GRAHAM 1976; RABINOWITZ et al. 1977; DIMMOCK and KENNEDY 1978; BARRETT et al. 1984b) or indistinguishable from titres in animals infected with only standard virus (DIMMOCK et al. 1986). These differences may reflect the quantity or quality of DI virus administered, since the death of animals with decreased levels of infectivity may merely be postponed by a few days or weeks. In addition, BARRETT et al. (1984b) have observed that the distribution of infectivity in protected mice varies with time depending on the DI virus isolate being studied, as the administration of one DI virus (DI SFV p13) resulted in reduced levels of infectivity throughout the period of infection, while with another DI virus (p4) infectivity levels were identical to those of mice inoculated with standard virus for the first 2 days of infection only, and declined thereafter. Interestingly, the amount of infectious SFV in one particular tissue (brain, olfactory lobes, spleen or serum) could be diminished by DI SFV p13 so that it was virtually spared infection, whereas the amount of virus in the other tissues was unaltered (BARRETT et al. 1984 b).

7.2 DI Virus

Attempts to detect DI virus in protected mice have proved unsuccessful in the majority of studies, but this may be due simply to a problem of sensitivity, since most in vitro DI assays used for in vivo work detect only $>10^6$ DI particles, with the honourable exception of POPESCU and LEHMANN-GRUBE (1977; see below). Nonetheless, it is surprising that DI virus cannot easily be found in animals co-inoculated with DI and standard viruses under conditions where one would expect DI virus to be propagated; i.e. normal amounts of infectivity are synthesized and there is no protection (BARRETT et al. 1984b; DIMMOCK et al. 1986). Why, it must be asked, is DI virus not propagated in vivo as in vitro? In fact, when investigating this problem HOLLAND and VILLARREAL (1975) found that DI VSV could be propagated only in neonatal mice, and that in adults it disappeared progressively on successive passages. They suggested that the sparsely myelinated neonatal brain resembled cells in culture more closely than did the adult brain and that cell-to-cell spread was facilitated. Indeed, virus antigen was present in most cells of neonatal brain, but only in scattered neurones in adults (HOLLAND et al. 1976b).

Isolation of DI virus from the brains of mice was made possible by the use of an amplification step in tissue culture (HOLLAND and VILLARREAL 1975; SPANDIDOS and GRAHAM 1976; DIMMOCK and KENNEDY 1978). Such procedures are too cumbersome for routine use, and a new methodology is badly needed. Taken at face value, these results suggest that virus multiplication in protected animals is controlled by low numbers of DI virus particles, but it is worth considering the possibility that if there are DI viruses which interfere in vitro but do not protect mice (BARRETT and DIMMOCK 1984b), it may be inappropriate to use the criterion of in vitro interference to identify DI virus isolated from animals. Possibly, DI virus which interferes in vivo is qualitatively different from that which interferes in vitro.

The only sensitive assay for DI virus is a 'reverse plaque assay', where DI virus provides a focus of cells protected from infection. It is available only for DI LCMV (POPESCU et al. 1976), DI VSV (WINSHIP and THACORE 1980) or DI rabies virus (KAWAI and MATSUMOTO 1982). Using this technique, the propagation of DI LCMV inoculated intracerebrally or intraperitoneally into both adult and newborn mice has been elegantly demonstrated (POPESCU and LEHMANN-GRUBE 1977).

On the basis of in vitro studies, HUANG (1977) suggested that DI and standard virus underwent a cyclic interaction whereby standard virus temporarily escaped inhibition by DI virus, and recently CAVE et al. (1985) have interpreted their in vivo data in this light. However, no evidence of cycling of DI and standard virus populations in vivo (e.g. SFV, BARRETT et al. 1984b) has been reported by others.

Recently, recombinant DNA technology has been applied to the detection of DI virus in vivo. CAVE et al. (1984, 1985) have prepared an 800-nucleotide complementary DNA (cDNA) copy of DI VSV RNA and used this as a probe to detect DI RNA in the brains of protected mice. Mice which had been co-inoculated with standard and DI virus were protected against death, and DI RNA could be detected in the brains of such mice. However, protected mice were always found to have standard virus RNA in addition. The technique was able to detect 2×10^7 or more genome equivalents of DI virus nucleic acid, but a quantitative comparison between detection of DI virus particles and DI virus nucleic acid was not reported. This would be of interest, as SOUTHERN et al. (1984) have used cDNA probes of LCMV to detect virus RNA in tissues of mice persistently infected with LCMV and noted a discrepancy between the level of viral nucleic acid and the titre of infectious virus. The excess of viral nucleic acid may be DI RNA, but this was not investigated. However in later experiments S.J. FRANCIS, M.K. SINGH, M.B.A. OLDSTONE and P.J. SOUTHERN (S.J. FRANCIS, personal communication) have examined the organs of such mice for viral RNA sequences using cDNA probes from the small (S) RNA genome segment. Northern blot analysis of RNA extracted from kidneys revealed not only genomic sized RNA and putative nucleoprotein and glycoprotein mRNAs but also a heterogeneous population of subgenomic RNA species. Further analysis of these subgenomic RNAs showed that the majority contained sequences from the 3' and extreme 5' ends of the S segment but only some contained sequences from central regions. Work is continuing to determine whether or not these subgenomic RNAs represent DI RNA species or are the products of aberrant transcription or processing.

Summary. DI virus can alter the distribution of standard virus, both qualitatively (in respect of the tissue in which it is found) and quantitatively. DI virus itself is usually present in curiously low amounts and has proved difficult to isolate from a variety of experimental systems. This may reflect merely the insensitivity of assays or more interesting possibilities. The situation is resolvable by the use of nucleic acid probes to detect DI genomes.

8 Changes in Pathology Resulting from Administration of DI Virus

As an additional measure of the effects of DI virus, histopathological changes in tissues of protected animals have also been determined. Standard VSV inoculated by the intracerebral route into adult mice causes an acute encephalitis resulting in death between 2 and 3 days post infection, while co-infection with DI VSV causes a more slowly progressive disease with death in 5–9 days and extensive pathology of the brain and spinal cord (DOYLE and HOLLAND 1973;

RABINOWITZ et al. 1977). The latter comprises histological changes not characteristic of the standard virus infection, including parenchymal necrosis in the spinal cord, demyelination in the white matter and spongiform changes in the grey matter (RABINOWITZ et al. 1977). DI reovirus acts similarly in 2-day-old rats co-inoculated intracerebrally with a lethal dose of standard virus. The majority of animals survive, but become runted (SPANDIDOS and GRAHAM 1976). No histopathological data were reported.

In contrast, brains of DI SFV-protected mice inoculated by the intranasal route showed no pathology or alterations whatsoever, and no virus antigen was detectable by biotin-streptavidin-labelled antibody system, although small quantities of standard virus were usually present (CROUCH et al. 1982; BARRETT et al. 1984b). However, when the same viruses were co-inoculated by the intraperitoneal route, a few of the mice developed a focal demyelination of a type usually seen during infection by avirulent strains of SFV, rather than the neurolysis characteristic of virulent SFV (T. ATKINSON, A.D.T. BARRETT, L. MCLAIN, A. MACKENZIE, and N.J. DIMMOCK, unpublished data; CROUCH et al. 1982).

Since SFV is a neurotropic virus, it was of interest to examine the neurotransmitter functions of the brains of DI SFV-treated mice. The changes in dopamine and serotonin metabolism which accompany the acute infection were diminished in mice protected by DI virus (A.J. CROSS, J.A. JOHNSON, A.D.T. BARRETT and N.J. DIMMOCK, unpublished data). However, clinically normal, virus-free survivors at 12 days post infection showed a very large (nearly tenfold) decrease in the activity of glutamate decarboxylase in the pathway of gamma-aminobutyric acid synthesis (BARRETT et al. 1986). Although this change was transient, it indicates how DI virus can modify the host in the absence of detectable infectious virus, antigen or histopathology.

Summary. DI viruses have the ability not only to prevent acute viral infections, but also to cause the pathology and the disease pattern to differ from those normally seen with the standard virus.

9 Persistent Infections

One of the original proposals of HUANG and BALTIMORE (1970) was that DI viruses were involved in the establishment and maintenance of persistent infections. The first studies to support this hypothesis were by SPANDIDOS and GRAHAM (1976), who showed that intracerebral infection of newborn rats with DI and standard reovirus resulted in survival of the rats, which developed a runting syndrome. Defective virus was isolated using an amplification method from brains of runted rats at 30 days post infection, but its interfering properties were not investigated. In the following year POPESCU and LEHMANN-GRUBE (1977) inoculated newborn mice with standard LCMV together with DI LCMV, which resulted in the generation and persistence of DI and infectious LCMV.

Both were present at 83 days post inoculation, but defective virus in particular was at a very low level. Another study by WELSH et al. (1977) showed that intracerebral co-inoculation of 2-day-old rats with DI and standard LCMV resulted in protection, but without persistence of virus in the animal. If a comparison were permissible, it could be argued that the presence of a high ratio of DI LCMV to LCMV biased the system for clearance, and that a lower ratio of DI LCMV to LCMV biased it for persistence, but the systems are too different to make this any more than a suggestion. Another view maintains that arenavirus persistence in vivo is controlled by immunoregulation [for a review see RAWLS et al. (1981)]. The isolation by JACOBSON and PFAU (1980) of a standard LCMV from persistently infected mice which is resistant to interference by DI LCMV can now be reinterpreted in the light of in vitro studies by HORODYSKI et al. (1983) and HORODYSKI and HOLLAND (1984), who have shown that standard virus in persistent infections evolved to a form resistant to interference by DI virus; this in turn generates its own DI virus, which then provides pressure on the new standard virus to evolve resistance against DI virus.

FULTZ et al. (1982b) have demonstrated that the acute infection caused by VSV in hamsters can become persistent following intraperitoneal inoculation with DI VSV. However, persistence could also be initiated by administration of poly I:poly C, and may be driven primarily by interferon, from whatever stimulus. Virus was isolated up to 8 months after inoculation, but only after co-cultivation for four passages. These were mutants of the original standard virus which were temperature-sensitive and gave small plaques. Further studies with this system suggested that the life expectancy of persistently infected hamsters was reduced (FULTZ et al. 1984), but the use of historical controls makes this uncertain. Histopathological lesions were also seen in the brains of persistently infected hamsters, while the same animals were shown to be immunocompetent, suggesting that persistent virus avoids immune surveillance. This is consistent with the detection of VSV antigens, but not infectious virus, in tissues at 13 months post infection (FULTZ et al. 1984). In contrast, studies using DI VSV in mice have shown protection but failed to show persistence, as determined by isolation of infectious virus and detection of DI or standard VSV RNA in the brains of protected animals (HOLLAND and VILLARREAL 1975; CAVE et al. 1985).

SFV is a virus which in mammals becomes persistent only when they are immunologically compromised (ATKINS et al. 1985). However, while the majority of mice which survived the lethal dose of virus through treatment by DI SFV were free of detectable infectious virus, a minority (approximately 16%) contained up to 10^5 pfu/brain. These viruses can be plaqued directly, and the isolation rate was not increased by blind passage. None were temperature-sensitive or plaque mutants. The infectious virus declined with time and was undetectable in the brain 89 days post infection. Mice from the same experimental group were free of histopathological lesions or virus antigen in the brain and spinal cord, but the actual mice from which virus was isolated were not examined. A particularly surprising discovery was that about one-third of the isolates were as virulent as the standard virus initially used to inoculate the mice, and some mice had the equivalent of 100 LD_{50} virus in persistently infected brain. How the virulence was contained (presumably by DI virus) is not known (ATKINSON et al. 1986).

A mirror image of what is usually meant by DI virus-mediated persistence is seen with a putative DI retrovirus, SFFV. When purified away from its helper LLV-F, the DI virus latently and persistently 'infects' cells. However, after 30 days it is converted to a form which can no longer be rescued by the helper (ECKNER and HETTRICK 1979).

Summary. In a number of systems, DI viruses facilitate the establishment of persistent infections which may be associated with a disease which differs from that caused by standard virus alone. DI virus can thus be added to the long list of factors which predispose to persistent infection.

10   DI Viruses and Tumorigenesis

A case can be made for regarding replication-defective tumour viruses which interfere with the replication of the helper virus as DI virus. However, against this view is the fact that genomes of defective retroviruses contain oncogene sequences which are derived from cellular DNA, and that defective virus production is not enhanced by helper virus as is that of conventional DI viruses.

Nonetheless, defective tumour virus particles can affect the process of tumorigenesis. One example involving the defective reticuloendotheliosis virus, REV-T, has already been discussed (see Sect. 6.4 and BOSE 1984). There, the defective virus is immunomodulatory and induces suppressor cells which act on a population of cytotoxic anti-tumour cells, potentiating the tumour. The opposite effect, inhibition of tumorigenesis, is exerted by DI particles of a parvovirus (adenoassociated virus) on adenovirus type 12 infection in newborn hamsters (KIRCHSTEIN et al. 1968; MAYOR et al. 1973; TOOLAN and LEDINKO 1968). Interestingly, the same is also true in tissue culture (DE LA MAZA and CARTER 1981), although DI adenoassociated virus will not prevent lytic replication of adenovirus in permissive cells (CARTER et al. 1979). Thus, we have another example where DI particles have different biological properties depending on the experimental situation.

O'Callaghan and co-workers have performed elegant studies on the DNA tumour virus equine herpesvirus 1 (EHV-1). DI particles do not contain the putative EHV-1 transforming sequences, which suggests that they do not play a direct role in the transformation process. Rather, it is thought that DI virus aids tumorigenesis by suppressing virus replication and preventing cytocidal events. Thus, the stable expresion and integration of transforming DNA sequences present in standard virus DNA is facilitated (BAUMANN et al. 1984).

To date there are no reports that co-inoculation of animals with DI and standard virus causes tumours. However, FLECKENSTEIN et al. (1975) report finding defective genomes of herpesvirus saimiri in post mortem tissues from virus-induced monkey tumours. Whether or not these genomes had interfering properties was not determined and the significance of this observation, if any, has yet to be appreciated.

Summary. Inasmuch as tumours are the result of virus–cell interactions, there seems to be every reason (and some evidence) to expect that DI virus can influence their development. DI virus could either enhance or inhibit tumorigenesis.

11 DI Viruses and Natural Infections

Although all viruses produce DI particles in the majority of tissue culture cells, and although some of these DI viruses have been shown to modulate infections of animals, there has been no unequivocal report of the isolation of DI virus from any natural virus infection, either of man or of any other animal. However, this failure is not surprising, as it has proved difficult to isolate DI viruses from experimental infections where they have been deliberately introduced and are demonstrably affecting virus multiplication and the course of disease (Sect. 7.2). Nonetheless, until DI virus is isolated from a natural infection its role in the expression of disease must be treated with caution.

Is there any evidence that DI particles may actually play a role in modulating a virus infection? ROBINSON (1978) remarked on purely hypothetical grounds that persistent infections caused by hepatitis B virus might involve DI virus and, more recently, RUIZ-OPAZO et al. (1982) suggested that while supercoiled DNA is present in infectious hepatitis B virions, relaxed circular DNA may be the form present in DI particles. Evidence has been sought, but not found, to implicate DI viruses in persistence of arenaviruses in small rodents (BUCHMEIER et al. 1980); infectious pancreatic necrosis virus in salmon with disease of that name (MacDONALD and YAMAMOTO 1978; DOBOS and ROBERTS 1983); and measles virus with subacute sclerosing panencephalitis (SSPE; HALL et al. 1974; CERNESCU and SORODOC 1980). In the last case, more recent work favours the view that SSPE is associated with defective synthesis of the matrix protein (CARTER et al. 1983).

More encouraging is the finding that rotaviruses isolated from two chronically infected immunodeficient children have an unusual genome structure, which is compatible with their being DI viruses (PEDLEY et al. 1984). Analysis of the genome by polyacrylamide gel electrophoresis has shown the presence of double-stranded RNA segments additional to the 11 segments normally present in rotavirus virions. The extra segments did not appear to result from an infection with more than one rotavirus, and hybridized to segment-specific recombinant DNA probes. It is of particular interest that some of the extra segments were considerably larger than the virion segment from which the probe was made, and appeared to be covalently linked concatomers, a frequent structural form of DI virus genomes of the Herpesviridae (FRENKEL 1981) and alphaviruses (LEHTOVAARA et al. 1981; MONROE and SCHLESINGER 1984). However, they were not regular MW multiples of virion segments. Concomitant with the extra segments was a decrease in the corresponding virion segment. Tissue culture studies with a rotavirus have shown that serial undiluted passage results in the isolation of virus particles with extra RNA segments similar to those isolated from the immunodeficient children (S. Pedley, personal communication). Thus, it is tempting to suggest that the two clinical isolates may contain DI particles. However, as human rotaviruses cannot be propagated in tissue culture it was not possible to demonstrate or investigate their interfering activity.

Another step forwards was the demonstration by BEAN et al. (1985) that an avirulent type A influenza virus isolated from chickens contained small "DI-type" RNAs, whereas a lethal form of the virus did not. They found no other differences in the antigenic or genomic structure of the isolates to explain the difference in virulence. One of the small RNAs was derived from the RNA segment encoding the P2 protein, as are most influenza DI RNAs (NAYAK et al. 1985). Evidence of interfering activity was shown by the reduction in pathogenicity in chickens infected with mixtures of the virulent strain and the avirulent, small RNA-containing strain. Overall data suggested that avirulence results from the presence of the small RNAs. However, the RNAs investigated were from viruses passaged twice through embryonated eggs before analysis and not directly from the chicken; hence it is possible that the small RNAs were synthesized by the avirulent strain during the isolation procedure. It is well known that influenza viruses differ markedly in the ease with which they generate DI virus (FAZEKAS DE ST. GROTH and GRAHAM 1954; MEIER-EWERT and DIMMOCK 1970). Nonetheless, the inference that DI RNAs may have reduced a mortality rate of 80% to one of around 2.6% should encourage further investigation.

Summary. The possibility that DI viruses exist in nature has been strengthened by recent studies with influenza and rotaviruses. To establish this beyond doubt we need to demonstrate the presence of characteristic DI genomes in the animal and to isolate a DI virus homologous with the relevant infecting standard virus.

12 DI Viruses as Antiviral Agents?

Because of their ability to inhibit replication of standard virus in vitro, DI viruses have been proposed as antiviral agents for prophylaxis in humans and other animals. This becomes more of a reality in view of the evidence presented above that DI viruses do indeed prevent or attenuate lethal infections in vivo. Currently there seems to be no possibility of using DI virus therapeutically, since it is usually necessary to co-inoculate the DI and standard virus to achieve protection. Possible contra-indications to the antiviral use of DI viruses is their involvement in persistent infections, modification of immune responses and neurochemical abnormalities. However, since there is extraordinary variation in the sequences of DI viruses – for example, DI alpha, influenza and vesicular stomatitis viruses (LEHTOVAARA et al. 1981, 1982; JENNINGS et al. 1983; NICHOL et al. 1984) – and nothing known of the relationship between genome sequence and biological function, it could well be that DI viruses having a particular genome sequence may prove to be satisfactory and safe antiviral agents. The biological diversity of DI viruses discussed in this review is perhaps not yet widely appreciated.

Summary. DI viruses have immense antiviral potential, but because of the prevailing ignorance of their biological properties their use could currently be sanctioned only in cases of accidental exposure to a lethal pathogen.

13 Conclusions

1. There is now no doubt that DI viruses have the ability to modulate infections in vivo, and the assumption that this ability resides in information encoded in their genomes has now been confirmed in some experimental systems.
2. Extreme forms of modulation such as the prevention of lethal infections and their conversion to chronic disease or long-term persistent infections with or without clinical signs, have been demonstrated. Changes in neurochemistry in clinically normal animals suggest that DI viruses are also capable of more subtle modulations.
3. It seems unlikely that interference with replication as described for in vitro systems is the sole mechanism by which DI viruses exert their effects in vivo, particularly since there are DI viruses which interfere well in vitro but have no effect in vivo (BARRETT and DIMMOCK 1984b).
4. Few generalizations can be made about how DI viruses modulate infections in vivo. This is not too surprising, since viruses differ widely in replicative strategy on the one hand and pathogenesis on the other. Variation in the biological properties of DI viruses themselves is another cause of variation.
5. We have argued that in vivo some DI viruses modify immune responses and that this could be relevant to the ways in which they modulate infection. If correct, this theory would also explain how relatively few DI virus particles exert their effects in a large multicellular organism. However, co-infection of a cell with standard virus is necessary for the propagation of DI virus, and we are not ruling out the importance of intracellular interference.
6. The genome(s) of DI viruses generated by a particular viral serotype show extensive sequence variation. This fact, together with the demonstration that a DI virus population contains extensive biological heterogeneity, shows that great care is needed in selecting a DI virus suitable for in vivo work. The difficulty of maintaining a constant and uniform population of DI viruses in passage may explain difficulties which have been experienced with in vivo systems. Bioassays have proved useful in checking for such variations.
7. It is strangely difficult to isolate DI virus from experimental in vivo infections in which virus multiplication and disease are evidently being modulated by DI virus. This may be the result of insensitive assay procedures or other unknown factors. The problem can be overcome by using molecular probes for DI genomic sequences.

8. In view of this, it is hardly surprising that DI viruses have not been isolated from "natural" infections. However, the physical and biological diversity of DI viruses, the quantitative nature of their interactions with their hosts, and their ability to affect the functions of the immune system suggest that they are capable of subverting normal infections perhaps into those with unusual or late-onset manifestations. The constant variability of DI viruses makes it likely that their effects will be idiosyncratic.

References

Ada GL, Leung KN, Ertl H (1981) An analysis of effector T cell generation and function in mice exposed to influenza A or Sendai viruses. Immunol Rev 58:5–24

Allison AC, Warwick RTT (1949) Quantitative observations on the olfactory system of the rabbit. Brain 72:186–197

Atkins GS, Sheahan BJ, Dimmock NJ (1985) Semliki Forest virus infection of mice: a model for genetic and molecular analysis of viral pathogenicity. J Gen Virol 66:395–408

Atkinson T, Barrett ADT, MacKenzie A, Dimmock NJ (1986) Persistence of virulent Semliki Forest virus in mouse brain following co-inoculation with defective interfering particles. J Gen Virol 67:1189–1194

Barrett ADT, Dimmock NJ (1984a) Variation in homotypic and heterotypic interference by defective interfering viruses derived from different strains of Semliki Forest virus and from Sindbis virus. J Gen Virol 65:1119–1122

Barrett ADT, Dimmock NJ (1984b) Modulation of Semliki Forest virus-induced infection of mice by defective interfering virus. J Infect Dis 150:98–104

Barrett ADT, Dimmock NJ (1984c) Properties of host and virus which influence defective interfering virus-mediated protection of mice against Semliki Forest virus lethal encephalitis. Arch Virol 81:185–188

Barrett ADT, Dimmock NJ (1984d) Modulation of a systemic Semliki Forest virus infection in mice by defective interfering virus. J Gen Virol 65:1827–1831

Barrett ADT, Crouch CF, Dimmock NJ (1981) Assay of defective-interfering Semliki Forest virus by the inhibition of synthesis of virus-specified RNAs. J Gen Virol 54:273–280

Barrett ADT, Crouch CF, Dimmock NJ (1984a) Defective interfering Semliki Forest virus populations are biologically and physically heterogeneous. J Gen Virol 65:1273–1283

Barrett ADT, Guest AR, Mackenzie A, Dimmock NJ (1984b) Protection of mice infected with a lethal dose of Semliki Forest virus by defective interfering virus: modulation of virus multiplication. J Gen Virol 65:1909–1920

Barrett ADT, Cross AJ, Crow TJ, Johnson JA, Guest AR, Dimmock NJ (1986) Subclinical infections in mice resulting from the modulation of a lethal dose of Semliki Forest virus with defective interfering viruses: neurochemical abnormalities in the central nervous system. J Gen Virol 67:in press

Baumann RP, Dauenhauer SA, Caughman GB, Staczek J, O'Callaghan DJ (1984) Structure and genetic complexity of the genomes of herpesvirus defective-interfering particles associated with oncogenic transformation and persistent infection. J Virol 50:13–21

Bean WJ, Kawaoka Y, Wood JM, Pearson JE, Webster RG (1985) Characterization of virulent and avirulent A/chicken/Pennsylvania/83 influenza A viruses: potential role of defective interfering RNAs in nature. J Virol 54:151–160

Bernkopf H (1950) Study of infectivity and haemagglutination of influenza virus in deembryonated eggs. J Immunol 65:571–583

Biron CA, Habu S, Okumura K, Welsh RM (1984) Lysis of uninfected and virus-infected cells in vivo: a rejection mechanism in addition to that mediated by natural killer cells. J Virol 50:698–707

Bose HR Jr (1984) Reticuloendotheliosis virus and disturbance in immune regulation. Microbiol Sci 1:107–112

Brinton MA (1983) Analysis of extracellular West Nile virus particles produced by cell cultures from genetically resistant and susceptible mice indicates enhanced amplification of defective interfering particles by resistant cultures. J Virol 46:860–870

Buchmeier MJ, Welsh RM, Dutko FJ, Oldstone MBA (1980) The virology and immunobiology of lymphocytic choriomeningitis virus infection. Adv Immunol 30:275–331

Campbell DE, Kemp MC, Perdue ML, Randall CC, Gentry GA (1976) Equine herpesvirus in vivo: cyclic production of a DNA density variant with repetitive sequences. Virology 69:737–750

Carter BJ, Laughlin CA, de la Maza LM, Myers M (1979) Adeno-associated virus autointerference. Virology 92:449–462

Carter MJ, Willcocks MM, ter Meulen V (1983) Defective translation of measles virus matrix protein in a subacute sclerosing panencephalitis cell line. Nature 305:153–155

Cave DR, Hagen FS, Palma EL, Huang AS (1984) Detection of vesicular stomatitis virus RNA and its defective-interfering particles in individual mouse brains. J Virol 50:86–91

Cave DR, Hendricksen FM, Huang AS (1985) Defective interfering particles modulate virulence. J Virol 55:366–373

Cernescu C, Sorodoc Y (1980) Subacute sclerosing panencephalitis and defective interfering measles virus particles. Rev Roum Med Virol 31:3–8

Crick J, Brown F (1977) In vivo interference in vesicular stomatitis virus infection. Infect Immun 15:354–359

Crouch CF, Mackenzie A, Dimmock NJ (1982) The effect of defective interfering Semliki Forest virus on the histopathology of infection with virulent Semliki Forest virus in mice. J Infect Dis 146:411–416

Darnell MB, Koprowski H (1974) Genetically determined resistance to infection with group B arboviruses. II. Increased production of interfering particles in cell cultures from resistant mice. J Infect Dis 129:248–256

de la Maza LM, Carter BJ (1981) Inhibition of adenovirus tumorigenicity by adenoassociated virus. J Natl Cancer Inst 67:1323–1326

Dimmock NJ, Kennedy SIT (1978) Prevention of death in Semliki Forest virus-infected mice by administration of defective-interfering Semliki Forest virus. J Gen Virol 39:231–242

Dimmock NJ, Beck S, McLain L (1986) Protection of mice from lethal influenza by DI virus: unsuspected mechanisms. J Gen Virol 67:839–850

Dobos P, Roberts TE (1983) The molecular biology of infectious pancreatic necrosis virus: a review. Can J Microbiol 29:377–384

Doyle M, Holland JJ (1973) Prophylaxis and immunisation in mice by use of virus-free defective T particles to protect against intracerebral infection by vesicular stomatitis virus. Proc Natl Acad Sci USA 70:2105–2108

Eckner RJ, Hettrick KL (1977) Defective spleen focus-forming virus: interfering properties and isolation free from standard leukaemia-inducing helper virus. J Virol 24:383–396

Eckner RJ, Hettrick KL (1979) Persistence and pathogenicity of defective Friend spleen focus-forming virus. J Exp Med 149:340–357

Faulkner G, Dubois-Dalcq M, Hooghe-Peters E, McFarland HF, Lazzarini RA (1979) Defective interfering particles modulate VSV infection of dissociated neuron cultures. Cell 17:979–991

Fazekas de St Groth S, Graham D (1954) The production of incomplete virus particles among influenza virus strains: experiments in eggs. Br J Exp Pathol 35:60–74

Fields S, Winter G (1982) Nucleotide sequence of influenza virus segments 1 and 3 reveal mosaic structure of a small viral RNA segment. Cell 28:303–313

Fleckenstein B, Bornkamm GS, Ludwig H (1975) Repetitive sequences in complete and defective genomes of *Herpesvirus saimiri*. J Virol 15:398–406

Frenkel N (1981) Defective interfering herpesviruses. In: Nahmias AJ, Dowdle WR, Schinazi RF (eds) The human herpes viruses. Elsevier/North-Holland, New York, pp 91–120

Fultz PN, Shadduck JA, Kang C-Y, Streilein JW (1981) On the mechanism of DI particle protection against lethal VSV infection in hamsters. In: Bishop DHL, Compans RW (eds) Replication of negative strand viruses. Elsevier/North-Holland, New York, pp 893–899

Fultz PN, Shadduck JA, Kang C-Y, Streilein JW (1982a) Mediators of protection against lethal systemic vesicular stomatitis virus infection in hamsters. Defective interfering particles, polyinosinate-polycytidylate and interferon. Infect Immun 37:679–686

Fultz PN, Shadduck JA, Kang C-Y, Streilein JW (1982b) Vesicular stomatitis virus can establish persistent infection in Syrian hamsters. J Gen Virol 63:493–497

Fultz PN, Holland JJ, Knobler R, Oldstone MBA (1984) Long term persistence by vesicular stomatitis virus in hamsters. In: Compans RW, Bishop DHL (eds) Nonsegmented negative strand viruses. Academic, New York, pp 489–496

Gamboa ET, Harter DH, Duffy PE, Hsu KC (1976) Murine influenza virus encephalomyelitis. III. Effect of defective interfering virus particles. Acta Neuropathol (Berl) 34:157–169

Hall WW, Martin SJ, Gould E (1974) Defective interfering particles produced during the replication of measles virus. Med Microbiol Immunol 160:155–164

Help GI, Coto CE (1980) Genesis de particulas interferentes durante la multiplication del virus Junin in vivo. Medicina (B Aires) 40:531–536

Hoelzer JD, Lewis RB, Wasmuth CF, Bose HR Jr (1980) Hematopoietic cell transformation by reticuloendotheliosis virus: characterization of the genetic defect. Virology 100:462–474

Holland JJ, Doyle M (1973) Attempts to detect homologous autointerference in vivo with influenza virus and vesicular stomatitis virus. Infect Immun 7:526–531

Holland JJ, Villarreal LP (1975) Purification of defective interfering T particles of vesicular stomatitis and rabies viruses generated in vivo in brains of newborn mice. Virology 67:438–449

Holland JJ, Villarreal LP, Breindl M (1976a) Factors involved in the generation and replication of rhabdovirus defective T particles. J Virol 17:805–815

Holland JJ, Villarreal LP, Breindl M, Semler BL, Kohne D (1976b) Defective interfering virus particles attenuate virus lethality in vivo and in vitro. In: Baltimore D, Huang AS, Fox CF (eds) Animal virology. Academic, New York, pp 773–786

Holland JJ, Semler BL, Jones C, Perrault J, Reid L, Roux L (1978) Role of DI virus, mutation and host response in persistent infections by enveloped RNA viruses. In: Stevens JG, Todaro GJ, Fox CF (eds) Persistent viruses. Academic, New York, pp 57–73

Holland JJ, Kennedy SIT, Semler BL, Jones CL, Roux L, Grabau E (1980) Defective interfering RNA viruses and the host cell response. In: Fraenkel-Conrat H, Wagner RR (eds) Comprehensive virology, vol 16. Plenum, New York, pp 137–192

Horodyski FM, Holland JJ (1984) Reconstruction experiments demonstrating selective effects of defective interfering particles on mixed populations of vesicular stomatitis virus. J Gen Virol 65:819–813

Horodyski FM, Nichol ST, Spindler KR, Holland JJ (1983) Properties of DI particle-resistant mutants of vesicular stomatitis virus isolated from persistent infections and from undiluted passages. Cell 33:801–810

Huang AS (1977) Viral pathogenesis and molecular biology. Bacteriol Rev 41:811–821

Huang AS, Baltimore D (1970) Defective viral particles and viral disease processes. Nature 226:325–327

Huang AS, Baltimore D (1977) Defective interfering animal viruses. In: Fraenkel-Conrat H, Wagner RR (eds) Comprehensive virology, vol 10. Plenum, New York, pp 73–116

Huang AS, Little SP, Oldstone MBA, Rao D (1978) Defective interfering particles: their effect on gene expression and replication of vesicular stomatitis virus. In: Stevens JG, Todaro GJ, Fox CF (eds) Persistent viruses. Academic, New York, pp 399–408

Hurd J, Heath RB (1975) Effect of cyclophosphamide on infections in mice caused by virulent and avirulent strains of influenza virus. Infect Immun 11:886–889

Jacobson S, Pfau CJ (1980) Viral pathogenesis and resistance to defective interfering particles. Nature 283:311–313

Jennings PA, Finch JT, Winter G, Robertson JS (1983) Does the higher order structure of the influenza virus ribonucleoprotein guide sequence rearrangements in influenza virus RNA? Cell 34:619–627

Jones CL, Holland JJ (1980) Requirements for DI particle prophylaxis in vesicular stomatitis virus infection in vivo. J Gen Virol 49:215–220

Kääriäinen L, Pettersson RF, Keränen S, Lehtovaara P, Söderlund H, Ukkonen P (1981) Multiple structurally related defective interfering RNAs formed during undiluted passages of Semliki Forest virus. Virology 113:686–697

Kawai A, Matsumoto S (1982) A sensitive bioassay system for detecting defective interfering particles of rabies virus. Virology 122:98–108

Kirchstein RL, Smith KO, Peters EA (1968) Inhibition of adenovirus 12 oncogenicity by adeno-associated virus. Proc Soc Exp Biol Med 128:670–673

Koprowski H (1954) Biological modification of rabies virus as a result of its adaptation to chicks and developing chick embryos. Bull WHO 10:709–724

Lehtovaara P, Söderlund H, Keränen S, Pettersson RF, Kääriäinen L (1981) 18S defective interfering RNA of Semliki Forest virus contains a triplicated linear repeat. Proc Natl Acad Sci USA 78:5353–5357

Lehtovaara P, Söderlund H, Keränen S, Pettersson RF, Kääriäinen L (1982) Extreme ends of the genome are conserved and rearranged in the defective interfering RNAs of Semliki Forest virus. J Mol Biol 156:731–748

Leung KN, Ada GL (1980) Cells mediating delayed-type hypersensitivity in the lungs of mice infected with an influenza A virus. Scand J Immunol 12:393–400

Liew FY, Russell SM (1983) Inhibition of pathogenic effect of effector T cells by specific suppressor T cells during influenza virus infection in mice. Nature 304:541–543

MacDonald RD, Yamamoto T (1978) Quantitative analysis of defective interfering particles in infectious pancreatic necrosis virus preparations. Arch Virol 57:77–89

Mayor HD, Houlditch GS, Mumford DM (1973) Influence of adenoassociated satellite virus on adenovirus-induced tumors in hamsters. Nature New Biol 241:44–46

Meier-Ewert H, Dimmock NJ (1970) The role of the neuraminidase of the infecting virus in the production of noninfectious (von Magnus) influenza virus. Virology 42:794–798

Mims CA (1956) Rift Valley fever virus in mice. IV. Incomplete virus: its production and properties. Br J Exp Pathol 37:129–143

Mims CA (1964) Aspects of the pathogenesis of virus diseases. Bact Rev 28:30–71

Monroe SS, Schlesinger S (1984) Common and distinct regions of defective interfering RNAs of Sindbis virus. J Virol 49:865–872

Monroe SS, Ou J-H, Rice CM, Schlesinger S, Strauss EG, Strauss JH (1982) Sequence analysis of cDNAs derived from the RNA of Sindbis virions and of defective interfering particles. J Virol 41:153–162

Nayak DP, Sivasubramanian N (1983) The structure of influenza virus defective interfering (DI) RNAs and their progenitor genes. In: Palese P, Kingsbury DW (eds) Genetics of influenza viruses. Springer, Vienna New York, pp 255–279

Nayak DP, Chambers TM, Akkina RK (1985) Defective-interfering (DI) RNAs of influenza viruses: origin, structure, expression and interference. Curr Top Microbiol Immunol 114:103–151

Nichol ST, O'Hara PJ, Holland JJ, Perrault J (1984) Structure and origin of a novel class of defective interfering particles of vesicular stomatitis virus. Nucleic Acids Res 12:2775–2790

O'Hara PJ, Nichol ST, Horodyski FM, Holland JJ (1984) Vesicular stomatitis virus defective interfering particles can contain extensive genome sequence rearrangements and base substitutions. Cell 36:915–924

Pedley S, Hundley F, Chrystie I, McCrae MA, Desselberger U (1984) The genomes of rotaviruses isolated from chronically infected immunodeficient children. J Gen Virol 65:1141–1150

Perrault J (1981) Origin and replication of defective interfering particles. Curr Top Microbiol Immunol 93:151–207

Popescu M, Lehmann-Grube F (1977) Defective interfering particles in mice infected with lymphocytic choriomeningitis virus. Virology 77:78–83

Popescu M, Schaefer H, Lehmann-Grube F (1976) Homologous interference of lymphocytic choriomeningitis virus: detection and measurement of interference focus-forming units. J Virol 20:1–8

Rabinowitz SG, Huprikar J (1979) The influence of defective interfering particles of the PR-8 strain of influenza A virus on the pathogenesis of pulmonary infection of mice. J Infect Dis 140:305–315

Rabinowitz SG, Dal Canto MC, Johnson TC (1977) Infection of the central nervous system produced by mixtures of defective-interfering particles and wildtype vesicular stomatitis virus in mice. J Infect Dis 136:59–74

Rawls WE, Chan MA, Gee SR (1981) Mechanisms of persistence in arenavirus infections: a brief review. Can J Microbiol 27:568–574

Robinson WS (1978) Persistent infection with hepatitis B virus. In: Stevens JG, Todaro GJ, Fox CF (eds) Persistent viruses. Academic, New York, pp 485–497

Roux L, Waldvogel FA (1983) Defective interfering particles of Sendai virus modulate HN expression on the surface of infected BHK cells. Virology 130:91–104

Roux L, Beffy P, Portner A (1984) Restriction of cell surface expression of Sendai virus HN glycoprotein correlates with its high instability in persistently and standard plus DI virus infected BHK-21 cells. Virology: 118–128

Roux L, Beffy P, Portner A (1985) Three variations in the cell surface expression of the haemagglutinin – neuraminidase glycoprotein of Sendai virus. J Gen Virol 66:987–1000

Ruiz-Opazo N, Chakraborty PR, Shafritz DA (1982) Evidence for supercoiled hepatitis B virus DNA in chimpanzee liver and serum dane particles: possible implications in persistent HBV infection. Cell 29:129–138

Rup BJ, Spence JL, Hoelzer JD, Lewis RB, Carpenter CR, Rubin AS, Bose HR Jr (1979) Immunosuppression induced by avian reticuloendotheliosis virus: mechanism of induction of the suppressor cell. J Immunol 123:1361–1370

Smith AL (1981) Genetic resistance to lethal flavivirus encephalitis: effects of host use and immune status and route of inoculation and production of interfering Banzi virus in vivo. Am J Trop Med Hyg 30:1319–1323

Southern PJ, Blount P, Oldstone MBA (1984) Analysis of persistent virus infections by in situ hybridization to whole-mouse sections. Nature 312:555–558

Spandidos DA, Graham AF (1976) Generation of defective virus after infection of newborn rats with reovirus. J Virol 20:234–237

Stark C, Kennedy SIT (1978) The generation and propagation of defective-interfering particles of Semliki Forest virus in different cell types. Virology 89:285–299

Sullivan JL, Mayner RE, Barry DW, Ennis FA (1976) Influenza virus infection in nude mice. J Infect Dis 133:91–94

Suzuki F, Ohya J, Ishida N (1974) Effect of anti-lymphocyte serum on influenza infection. Proc Soc Exp Biol Med 146:78–84

Toolan HW, Ledinko N (1968) Inhibition by H-1 virus of the incidence of tumors produced by adenovirus type 12 in hamsters. Virology 35:475–478

von Magnus P (1951) Propagation of the PR8 strain of influenza A virus in chick embryos. III. Properties of the incomplete virus produced in several passages of undiluted virus. Acta Pathol Microbiol Scand 29:157–181

Walker MH, Rup BH, Rubin AS, Bose HR Jr (1983) Specificity in the immunosuppression induced by avian reticuloendotheliosis virus. Infect Immun 40:225–235

Welsh RM, Oldstone MBA (1977) Inhibition of immunologic injury of cultured cells infected with lymphocytic choriomeningitis virus: role of defective interfering virus in regulating viral antigenic expression. J Exp Med 145:1449–1468

Welsh RM, Lampert PW, Oldstone MBA (1977) Prevention of virus-induced cerebellar disease by defective interfering lymphocytic choriomeningitis virus. J Infect Dis 136:391–399

Wiktor TJ, Dietzschold B, Leamnson RN, Koprowski H (1977) Induction and biological properties of defective interfering particles of rabies virus. J Virol 21:626–635

Winship TR, Thacore HR (1980) A sensitive method for quantification of vesicular stomatitis virus defective interfering particles: focus-forming assay. J Gen Virol 48:237–240

Wunner WH, Clark HF (1980) Regeneration of DI particles of virulent and attenuated rabies virus: genome characterization and lack of correlation with virulence phenotype. J Gen Virol 51:69–81

Wyde PR, Couch RB, Mackler T, Cate TR, Levy BM (1977) Effects of low- and high-passage influenza in normal and nude mice. Infect Immun 15:221–229

The Regulation of Lymphocyte Traffic

E.C. BUTCHER

1 Introduction: The Non Random Distribution of Functional Lymphoid Subsets In Vivo

The distribution of lymphocytes among the tissues of the body is not random. Both the number and the representation of particular functional subsets of lymphocytes are carefully controlled, differing in each lymphoid organ or tissue in a manner that presumably reflects local immune requirements. The tissue distribution of lymphocytes is a function of lymphocyte *class*, of their *previous history* or *stage of differentiation*, and of their *antigenic specificity*. For example, those lymphocyte populations primarily responsible for humoral immunity (B cells) predominate in the spleen and in the gut-associated Peyer's patches, where-

Department of Pathology, Stanford University Medical Center, Stanford, CA 94305, USA and The Veterans Administration Medical Center, Palo Alto, CA, USA

Current Topics in Microbiology and Immunology, Vol. 128
© Springer-Verlag Berlin·Heidelberg 1986

as T cells, which are primarily regulatory and cytotoxic cells, are the major lymphocyte type in the peripheral lymph nodes and skin (STEVENS et al. 1982; STREILEIN 1978). Functionally and antigenically defined T-cell subsets are also unequally distributed between mucosal and nonmucosal lymphoid tissues (ELSON et al. 1979; KRAAL et al. 1983). The distribution of certain effector and effector-precursor populations can be even more restricted: especially dramatic is the segregation of IgA- vs. IgG-expressing B cells. Surface IgA-bearing lymphocytes are highly represented in the mucosa-associated lymphoid organs, and the mucosal surfaces attract predominantly IgA-secreting plasma cells (GUY-GRAND et al. 1974; reviewed by LAMM 1976). In nonmucosal sites, such as peripheral lymph nodes or the skin, IgA-bearing cells are rare, and most plasma cells secrete IgM or IgG. Lymphocytes can also segregate in vivo on the basis of antigen specificity: antigen-specific B and T cells are disproportionately represented in lymph nodes or spleen challenged with antigen (KRAAL et al. 1982; SPRENT 1980) and antigen-specific plasma cells accumulate in tissue sites of specific antigen deposition (HUSBAND and GOWANS 1978; HUSBAND 1982).

These examples illustrate the existence of mechanisms directing and restricting the tissue distribution of functionally important subsets. This review is intended to outline our current knowledge about the nature of these mechanisms and to synthesize this knowledge into a coherent, if occasionally speculative, picture of the control of the migration and tissue localization of lymphocytes throughout their life cycle.

2   Regulation of the Distribution of Lymphocyte Subsets by Directed Migration: Role of Selective Lymphocyte-Endothelial Cell Recognition

2.1   Developmental Regulation and Organ-Specificity of Lymphocyte Interactions with High Endothelial Venules (HEV)

Most mature lymphocytes continuously circulate between the various lymphoid organs and other tissues of the body, traveling via the lymph and bloodstream. These lymphocytes are said to recirculate because they move from the bloodstream into lymphoid organs, then to the collecting efferent lymphatics, and eventually back to the bloodstream where they reenter the cycle (reviewed in FORD 1975; SPRENT 1977). Although the pace of recirculation is a function of lymphocyte class and stage of differentiation, the average lymphocyte completes this recirculatory cycle, and thus finds itself in a new lymphoid organ or tissue, roughly every 1–2 days. This remarkable cellular shuffling allows the full repertoire of lymphocyte specificities to be available for immune reaction throughout the body and probably also facilitates the cell-cell interactions required for the generation and control of immune responses.

Essential to this process of recirculation is the ability of migrating lymphocytes to leave the blood at appropriate sites. Lymphocytes have the remarkable capacity to recognize and bind selectively to specialized endothelial cells in lymphoid organs and sites of inflammation, binding initially to the luminal surface and then migrating through the vessel wall into the surrounding tissues. Outside of the spleen, most such migration occurs through the postcapillary venules in lymph nodes and Peyer's patches (GOWANS and KNIGHT 1964; MARCHESI and GOWANS 1964; see Fig. 1). These vessels are characterized by distinctive plump endothelial cells and thus are referred to as "high endothelial venules," or HEV. The interaction of lymphocytes with HEV is of central importance in controlling lymphocyte traffic and has been studied extensively using an in vitro model, first developed by STAMPER and WOODRUFF (1976), in which viable lymphocytes recognize and bind to HEV in frozen sections of murine or human (JALKANEN and BUTCHER 1985) lymph nodes or mucosal lymphoid organs (see Fig. 4, for example). The in vitro binding of lymphocyte populations to HEV in frozen sections accurately reflects their capacity to adhere to HEV under physiologic conditions (BUTCHER et al. 1979).

Results obtained with this assay, and confirmed in short-term in vivo homing experiments, demonstrate that lymphocyte-HEV recognition is not only an essential determinant of the ability of individual lymphocytes to recirculate, but also selectively directs the traffic of particular lymphocyte subsets through specific organs or body regions. Surface receptor systems exist permitting lymphocytes to discriminate between HEV in peripheral lymph nodes, in mucosa-associated lymphoid tissues (Peyer's patches or appendix), in inflamed synovium, and probably in other organs or tissues as well. These mechanisms of organ-specific HEV recognition control lymphocyte distribution to particular tissues. The regulation of the capacity of lymphocytes to recognize peripheral node vs. mucosal HEV during lymphocyte development and differentiation is discussed below.

2.1.1 Migratory Properties of Mature "Virgin" Lymphocytes

The ability to recognize and bind to HEV is a developmentally acquired property of mature lymphocytes. Mature B and T lymphocytes bind much better to HEV than do their immature precursors in the bone marrow and thymus (BUTCHER et al. 1979). In fact, the expression of functional levels of receptors for endothelial cells by lymphocytes may be closely linked with the migration of lymphocytes from the primary lymphoid organs to the periphery, where they join the recirculating population. As early as 30 min after local labeling of cells in the thymus, thymic migrant cells can be identified within the HEV-bearing organs (SCOLLAY et al. 1980). It is interesting to speculate that the expression of the ability to interact with endothelial cells may in some manner determine, or at least be developmentally linked to, the elements responsible for immigration of newly matured lymphocytes to the periphery.

These newly matured lymphocytes, and in fact the vast majority of peripheral lymphocytes, have never encountered specific antigen. These "virgin" lymphocytes are largely competent to bind HEV and circulate, but they exhibit endothelial interaction specificities that are developmentally predetermined, based on lymphocyte class (STEVENS et al. 1982; KRAAL et al. 1983). The greatest differ-

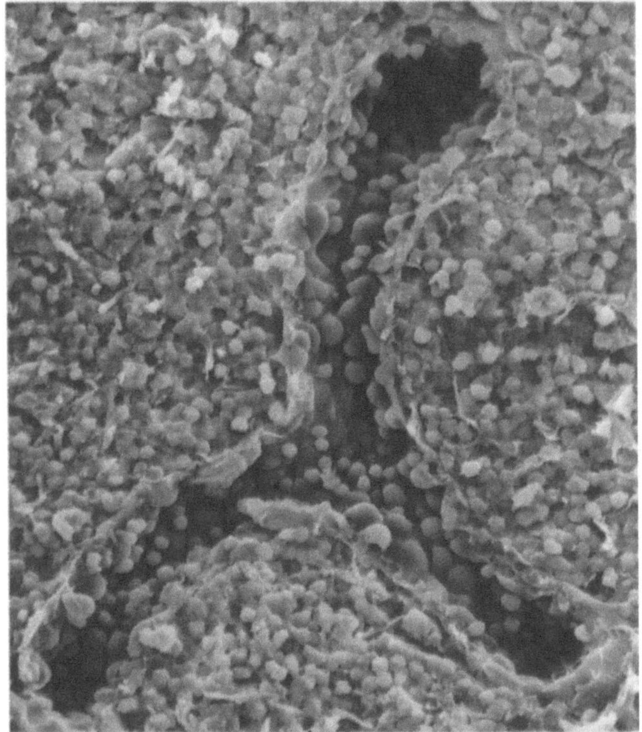

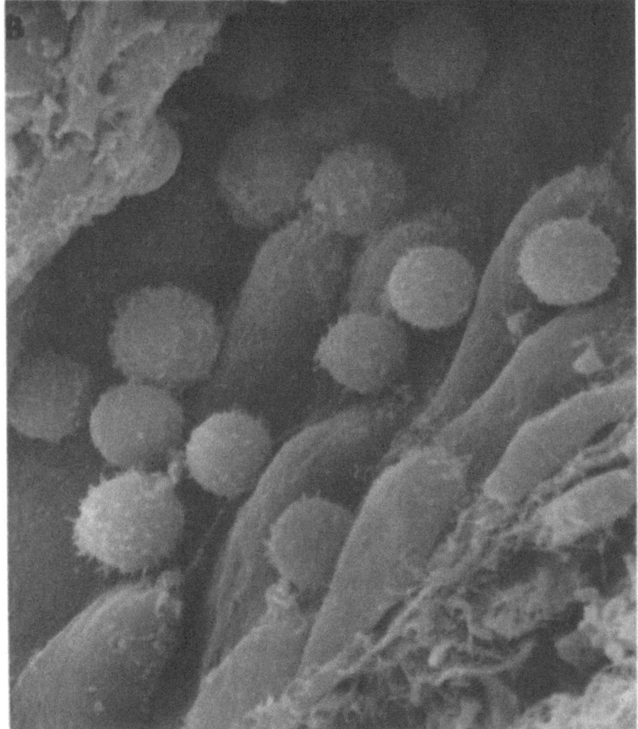

Fig. 1A, B. Scanning electron micrographs illustrating high endothelial venules (HEV) in mouse lymph nodes. **A** Inverted Y-shaped HEV. The high endothelial cells bulge into the lumen. Individual small round lymphocytes can be seen attached to the endothelium, some apparently migrating between endothelial cells on their way into the lymph node parenchyma. **B** Higher power showing numerous lymphocytes tightly bound to the high endothelial cells. Nonadherent blood elements were washed out by perfusion of the vasculature with medium prior to fixation by perfusion with glutaraldehyde in PBS. (From BUTCHER et al. 1982b, with permission)

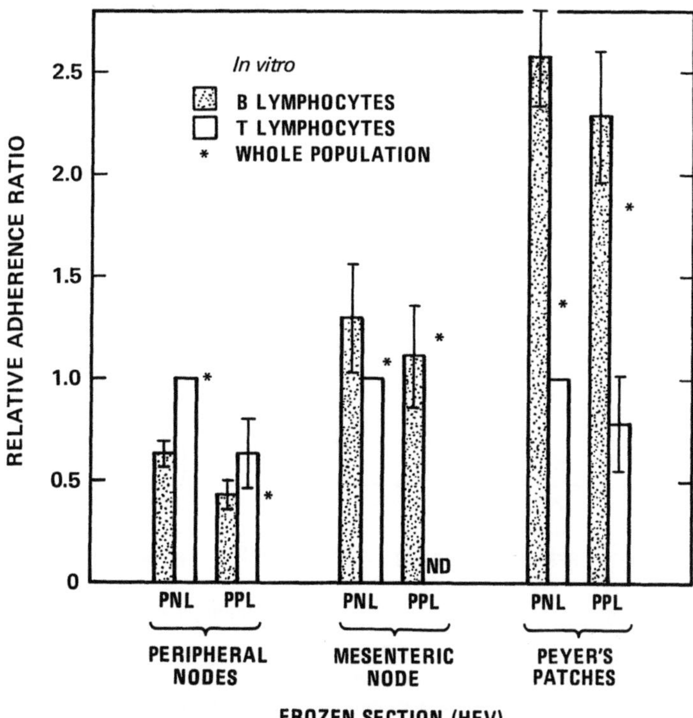

Fig. 2. Relative ability of B- and T-lymphocyte populations to bind in vitro to HEV. Adherence of peripheral node (*PNL*) and Peyer's patch lymphocyte (*PPL*) populations was assayed on HEV in frozen sections of peripheral (axillary and branchial) or mesenteric lymph nodes, or Peyer's patches. The relative adherence ratio (*RAR*) is the calculated number of sample cells binding to HEV per PN T cell bound under the same conditions. Error bars represent standard errors. (From STEVENS et al. 1982, with permission)

ences are found in comparisons of T- vs. B-lymphocyte populations. As shown in Fig. 2, T lymphocytes bind better than B cells to peripheral node HEV (about 1.5 ×), whereas B cells bind 2–3 times as well as T cells to HEV in the mucosal lymphoid organs (Peyer's patches). Interestingly, both populations bind equally well to HEV in the mesenteric lymph node. These different interactions with HEV are reflected in, and presumably determine, a parallel differential localization of B- and T-lymphocyte populations in these organs 2 h after i.v. injection (Fig. 3). T lymphocytes localize much better than B cells in peripheral lymph nodes and slightly less well than B lymphocytes in Peyer's patches. Again, the mesenteric node is intermediate in character. Localization in the spleen will be discussed below. The HEV binding and in vivo migratory characteristics of these unstimulated B and T cells are independent of the organ source of the sample lymphocytes (e.g., peripheral node B cells and Peyer's patch B cells exhibit the same characteristic endothelial and localization preferences), suggesting that *lymphocyte class is a major and perhaps the sole determinant of the migratory properties of mature but "virgin" lymphocytes.* Differential migration

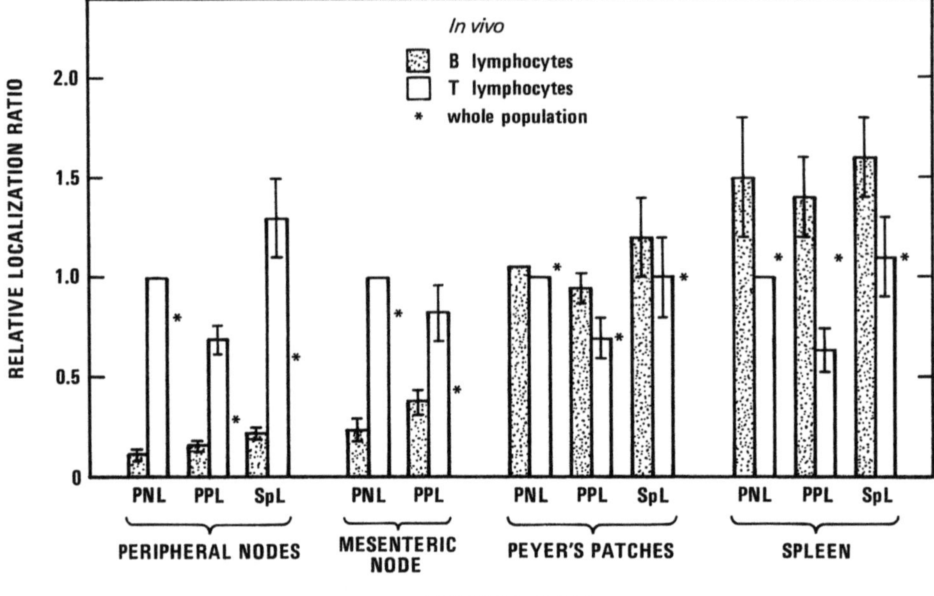

Fig. 3. Relative localization of B (Ig$^+$) and T (Thy-1$^+$) lymphocyte populations in various organs 2 h after i.v. injection. The relative localization ratio (RLR) is the calculated number of sample cells that would localize in a recipient organ per peripheral node (PN) T cell localizing if an equal number of each were injected. (PN T cells are an arbitrarily selected reference population.) Error bars represent standard errors. PNL, peripheral node lymphocytes; PPL, Peyer's patch lymphocytes; SpL, spleen cells. (From STEVENS et al. 1982, with permission)

of B- vs. T-cell populations has recently been confirmed in studies of lymphocyte homing in the guinea pig (VAN DINTHER-JANSSEN et al. 1983) and of lymphocyte traffic in rats (FOSSUM et al. 1983).

Smaller but still significant differences in endothelial and migration specificities are exhibited by the two major subpopulations of T-lymphocytes (KRAAL et al. 1983), distinguished in the mouse by their expression of surface Lyt-2 antigen. Regardless of their source, Lyt-2$^+$ cells (suppressor/killer phenotype) bind more frequently than Lyt-2$^-$ (helper/inducer phenotype) T cells (about 1.5 times) to peripheral node HEV, whereas both populations bind equally well to HEV in Peyer's patches. Again, these endothelial preferences appear to determine a parallel differential localization in vivo, with Lyt-2$^-$ and Lyt-2$^+$ T cells demonstrating slight relative preferences for localizing in Peyer's patches and peripheral lymph nodes, respectively.

The relative preference of B cells (vs. T cells) and of Lyt-2$^-$ T cells (vs. Lyt-2$^+$ T cells) for Peyer's patches over peripheral nodes may reflect a greater evolutionary requirement for precursors of, and help for, humoral immunity in mucosal sites. Regardless of such speculation, however, the results clearly suggest that *selective lymphocyte migration helps determine the relative availability of functionally distinct "virgin" lymphocyte populations in mucosal vs. nonmucosal organs*. This suggestion is supported by consideration of the phenotype

Table 1. Differences in the representation of B- and T-lymphocytes in various lymphoid organs[a]. (From STEVENS et al. 1982,
with permission)

Organ	Ig[+] (%)	Thy-1[+] (%)
Peripheral lymph nodes[b]	25 (4)	67 (5)
Mesenteric lymph node	38 (4)	59 (4)
Peyer's patches	76 (5)	13 (1)
Spleen	56 (5)	26 (5)

[a] Cell suspensions, stained for surface Ig or Thy-1, were analyzed by FACS. Each value is the mean of determinations on
eight male mice, ranging from 5 weeks to 7 months old; standard deviations are in parentheses
[b] Axillary and brachial nodes

of lymphocytes found in situ in these organs. The representation of B cells
(Table 1) and of Lyt-2$^-$ and Lyt-2$^+$ T cells in the major lymphoid organs
parallels the ability of the organs to extract these lymphocyte subsets from
the blood (STEVENS et al. 1982; KRAAL et al. 1983).

In Summary, mature but virgin lymphocyte populations exhibit developmentally predetermined organ-selectivity of HEV recognition, migration and tissue
distribution. However, during subsequent phases in lymphocyte differentiation,
which are initiated by activation to blastogenesis, the migration properties of
lymphocytes are specifically modulated. Lymphocyte activation occurs normally
as a result of lymphocyte interactions with specific antigen in the appropriate
cellular context (i.e., in the presence of antigen-presenting cells, regulatory T
cells, etc.) and abnormally in the case of lymphoid malignancies.

2.1.2 Effect of Lymphocyte Activation

In murine systems, activation to blastogenesis usually results in suppression
of migratory phenotype.

In Vitro Activated T Cells. Cloned T cells, maintained in culture with antigen
and a source of T-cell growth factors, represent an easily studied source of
normal or quasinormal activated T-lymphocytes. Such clones maintain many
features of the activated T cells from which they are derived, including antigen
specificity, H-2 restriction, cytotoxicity, etc. and therefore it might be expected
that their migratory properties are also representative of certain normal activated T cells in vivo. Thus, it is intriguing that cloned murine T cells, regardless
of antigen specificity, function, or source, uniformly fail to recognize or bind
to HEV in vitro (Table 2) and are unable to migrate efficiently into the HEV-
bearing organs after i.v. injection (DAILEY et al. 1982). These findings, observed
in studies of many different murine T-cell clones and lines, suggest that T cells
may normally pass through a stage (perhaps that stage which is most sensitive

Table 2. Homing receptor expression and HEV-binding properties of selected mouse and human lymphoid cells

A. Mouse cells	MEL-14	Specificity	Relative adherence to HEV[a]		
			Peripheral node	Mucosal	Synovial
Lymph node cells	+ +	Multiple	unity	unity	unity
38C13(B)	+ + +	Lymph node	1.1	0[b]	?
BK37(B)	+ + +	Lymph node	0.7	0	?
TK1(T)	–	Mucosal	0.1	5	?
TK38(T)	+ +	Dual	0.6	0.4	?
RAW112 (pre-B)	–	Non-binding	0	0	?
Thymocytes	±		0.1	0.1	?
Germinal center cells	–		0	0	?
Gut intraepithelial leukocytes	–	Mucosal	0	1.1	?
MEL-14 Peyer's patch T cells	–	Mucosal	0	0.9	?
Mesenteric node blasts		Mucosal	0	1.5	?
T cell clones	–	Non-binding	0	0	?

B. Human cells	Hermes-1	Specificity	Relative adherence to HEV[a]		
			Peripheral node	Mucosal	Synovial
Peripheral blood lymphs	+ +	Multiple	unity	unity	unity
LB-25(B)	+ + +	Lymph node	0.8	0[c]	0
KCA(B)	+ + +	Mucosal	0	0.8	0
IBW4(B)	+ + +	Dual	0.6	0.8	?
KW(B)	–	Non-binding	0	0	0
Thymocytes	±		0.2	?	0.2
Lamina propria blasts	+ + +		0	4–5	?
T cell clones	+ + +		0.7–4	1–3	?

[a] The relative adherence ratios given represent the calculated number of sample lymphocytes that would bind to HEV per control mouse lymph node cell (in A) or peripheral blood lymphocyte (in B) bound under identical conditions. Binding of the control populations defines an RAR of unity. Values given are selected from representative experiments
[b] Relative adherence ratio ≤ 0.05
[c] Relative adherence ratio ≤ 0.1

to the mitogenic influences of IL-2) in which they are not competent to migrate. Studies of T-cell blasts generated in mixed leukocyte cultures, or induced by mitogenic stimulation with concanavalin A, reveal that blastogenesis usually results in a rapid reduction and loss of HEV-binding ability, probably within hours and certainly by 1–2 days after stimulation (DAILEY et al. 1983). These findings, based exclusively on studies of blasts maintained in vitro, must obviously be interpreted with caution. However, the existence in vivo of a transient, nonmigratory stage following T-cell activation is supported by the work of SPRENT (1980), who showed that antigen-activated T cells are selectively retained in the spleen for 1–3 days following i.v. antigen administration.

Recent studies indicate that the regulation of receptors for HEV is even more complex, with differing responses to submitogenic vs. mitogenic activation

signals (HAMANN et al. 1986). The nondividing fraction of lymphocytes in mito-
gen-activated cultures, as well as in 3–5 day autologous mixed lymphocyte cultures, express 2–4 times increased levels of antigenically defined receptors for
HEV (identified by monoclonal antibody MEL-14 – see below). In addition,
HAMANN et al. (1986) confirmed that blastogenesis results in suppression of
HEV receptor expression on most dividing cells, but described a significant
subset of blasts, present following most mitogenic stimuli, that retained high
levels of antigenically defined receptors. Whether these receptor-bearing blasts
represent a distinct subset, or instead a transient stage in the activation process,
is uncertain.

HAMANN et al. (1984) also showed that mitogenesis results in induction of
another adherence system in which the gp180/95 LFA-1 complex plays an important role, leading to self-aggregation of dividing blasts. It seems likely that
the suppression of HEV-binding ability may be regulated coordinately with
the induction of other cellular adhesion mechanisms responsible for the arrest
of lymphocytes initially responding to antigen.

The regulation of migratory properties in response to lymphocyte activation
may be different in the human. Although the HEV-binding ability of mitogen-
stimulated lymphoblasts has not been studied, human T-cell clones, in contrast
to their mouse counterparts, uniformly bind well to HEV. This may reflect
differences in the regulation of receptors for HEV in humans, or perhaps may
indicate that IL-2 responsive T-cell clones represent a somewhat different stage
of normal T-cell differentiation in the two species. In either case, their expression
of functional HEV-binding capacity, which is required for normal lymphocyte
migration, raises the possibility that human T-cell clones may be considerably
better at mediating systemic immune responses than is suggested by murine
models.

Germinal-Center B Cells. Within the B-cell lineage, the earliest population of
in vivo antigen-activated cells that can be isolated readily for critical analysis
is the population of germinal-center lymphocytes. Germinal centers are discrete,
histologically defined foci of lymphoblasts that arise in response to antigenic
stimulation, and that probably provide a microenvironment important in heavy
chain class switching (KRAAL et al. 1982; BUTCHER et al. 1982c) and in the
differentiation of memory B cells (THORBECKE et al. 1974; KLAUS and KUNKL
1981). Germinal-center lymphoblasts are a phenotypically unique population
of mature B cells that can be defined and studied by virtue of their characteristically intense staining with the beta-galactosyl-specific lectin peanut agglutinin
(PNA; ROSE et al. 1980; BUTCHER et al. 1982c) and their absence of surface
IgD (BUTCHER et al. 1982c). They express low levels of surface immunoglobulin,
and on many germinal-center cells this immunoglobulin is demonstrably specific
for the inducing antigen (KRAAL et al. 1982). On cells in long-term stimulated
germinal centers in peripheral nodes, the predominant surface immunoglobulin
is IgG (KRAAL et al. 1982), while IgA predominates in Peyer's patches (BUTCHER
et al. 1982c), suggesting that these cells are precursors of the IgG- and IgA-
secreting plasma cells that characterize humoral immunity in these sites. Furthermore, germinal centers may be the major site of generation of memory B cells.

Germinal-center cells from lymph nodes are highly efficient in transferring antigen-specific memory responses to adoptive recipients. After primary stimulation, most transferable memory is contained in the germinal-center population (COICO et al. 1983; KRAAL et al. 1985). After secondary stimulation, B-cell memory is divided between the PNA^{hi} (germinal-center cell) and PNA^{lo} B-cell fractions, suggesting that with continued stimulation germinal centers may seed the body with memory cells (COICO et al. 1983; KRAAL et al. 1985).

B-cells do not express functional receptors for endothelium while in the germinal center microenvironment. Germinal center cells fail to recognize and bind HEV in vitro, and they are not competent to migrate in significant numbers from the blood into HEV-bearing organs in vivo (REICHERT et al. 1983). As precursors of specific memory and plasma cells, however, germinal-center lymphocytes must give rise to cells that are able eventually to migrate selectively to mucosal or nonmucosal sites (in order to explain the selective localization of IgG and IgA plasma cells in these tissues). In fact, preliminary studies with a monoclonal antibody to a lymphocyte receptor for HEV (see Sect. 2.2.2) indicate that many of the PNA^{lo} memory cells that arise after secondary (or long-term primary) immunization bear surface "homing receptors" and thus may well be competent to join the circulating lymphocyte pool (G. KRAAL et al. 1985).

2.1.3 Organ-Specific Lymphocyte-HEV Recognition Mechanisms

Malignant lymphomas represent an easily accessible population of activated lymphocytes, and their study has given significant insight into the endothelial interaction and migration specificities possible for their normal counterparts (BUTCHER and WEISSMAN 1979; BUTCHER et al. 1980). As illustrated in Fig. 4 and Table 2, certain lymphomas bind only to HEV in peripheral lymph nodes, and others bind almost exclusively to HEV in the mucosa-associated Peyer's patches. Both peripheral node and Peyer's patch HEV-specific lymphocytes bind to mesenteric node HEV (not shown). Another major group of lymphomas fails to bind to any type of HEV. This spectrum of endothelial interaction abilities of lymphomas is of obvious potential significance in relation to their patterns of spread and growth. In fact, preliminary studies of the growth patterns mouse lymphomas passaged subcutaneously indicate that HEV-binding lymphomas tend to exhibit early hematogenous spread with generalized symmetric involvement of lymphoid organs, whereas nonbinding lymphomas are characterized predominantly by local growth at the site of injection, with local lymphatic metastases (R. BARGATZE, N. WU, I.L. WEISSMAN, E.C. BUTCHER, manuscript in preparation).

Initial analyses of the expression of antibody-defined receptors for HEV (see Sect. 2.2.2) by human lymphoid neoplasms also demonstrate a correlation with dissemination to multiple lymphoid sites.

The ability of certain lymphomas to bind exclusively to HEV in peripheral nodes, and of others to bind to HEV in Peyer's patches, demonstrated the existence of "receptor" mechanisms permitting nearly absolute discrimination

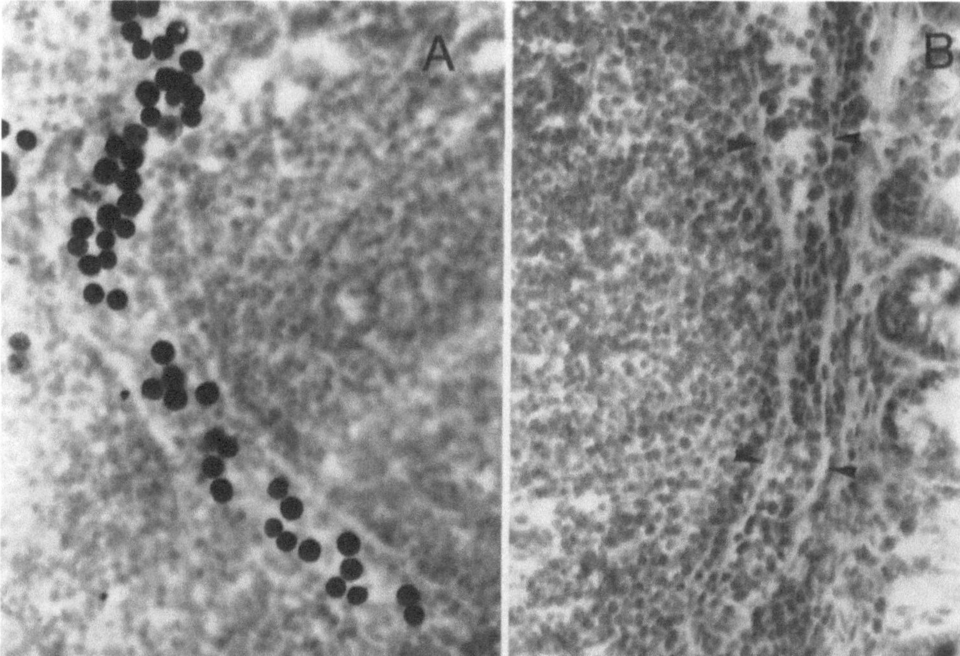

Fig. 4A, B. Highly selective binding by a lymph node-specific lymphoma. Lymphoma cells were incubated simultaneously on frozen sections of peripheral nodes and Peyer's patches. **A** Numerous cells bound to peripheral node HEV. Adherent cells are easily visualized because they stain more heavily than the underlying fragmented cells in the tissue section. **B** No cells on HEV in adjacent section of Peyer's patch. (From BUTCHER et al. 1982b, with permission)

between mucosal and nonmucosal high endothelium. The following sections illustrate the use of these organ-specific homing mechanisms by normal lymphocyte-effector or effector-precursor populations.

2.1.4 Organ-Specific Endothelial Cell Recognition by Some Differentiated Immunoblasts and Lymphocytes

Following local differentiation, some normal lymphoblasts and lymphocytes are induced or selected for expression of organ-specific homing receptors.

Mature (Migrating) Immunoblast Populations. In contrast to the nonmigratory germinal-center cells and T-cell clones, it is well documented that some populations of normal dividing lymphocytes are capable of migrating efficiently in vivo. For instance, stimulated peripheral lymph nodes contain a population of immunoblasts that localize selectively in lymph nodes (GUY-GRAND et al. 1974; GRISCELLI et al. 1969; MCWILLIAMS et al. 1975; GUY-GRAND et al. 1978; SMITH et al. 1980; MCDERMOTT and BIENENSTOCK 1979; HALL et al. 1979; ROSE et al. 1976) and peripheral sites of inflammation (ROSE et al. 1976) after i.v.

transfer, largely avoiding the mucosal organs and Peyer's patches. The specificity of migration is reversed in lymphoblasts derived from antigenic stimulation of mucosal surfaces. Although IUdR-labeled Peyer's patch lymphocytes (which consist largely of germinal-center cells) demonstrate only a meager migratory capability (GUY-GRAND et al. 1974; McWILLIAMS et al. 1975), dividing Peyer's patch cells eventually give rise to a population of dividing cells in the draining mesenteric lymph node and thoracic duct (HUSBAND et al. 1977) that migrate well in vivo, and localize specifically in the gut wall and in Peyer's patches (GUY-GRAND et al. 1974; GRISCELLI et al. 1969; McDERMOTT and BIENENSTOCK 1979; HALL et al. 1979; ROSE et al. 1976; HUSBAND et al. 1977). In contrast to the subset-dependent, source-independent migratory preferences of presumed "virgin" lymphocyte populations, the specificity of migrating immunoblasts appears to be source dependent and subset independent – for instance, both B and T mesenteric node immunoblasts can migrate specifically to the mucosal tissues or Peyer's patches (GUY-GRAND et al. 1974; McWILLIAMS et al. 1975; GUY-GRAND et al. 1978). The differential migratory patterns of peripheral node and gut-derived immunoblasts are well illustrated by the work of SMITH et al. (1980) presented in Fig. 5.

As might have been predicted from the HEV-binding characteristics of lymphomas, the selective migration of these normal lymphoblasts appears to be determined by selective recognition of organ-specific endothelial determinants. Gut-derived immunoblasts in mouse mesenteric lymph nodes (BUTCHER et al. 1982a) or blasts isolated from human lamina propria (S. JALKANEN and R. MacDERMOTT, personal communication) bind almost exclusively to HEV in Peyer's patch frozen sections, adhering very poorly to those in peripheral lymph nodes (Table 2). In vitro studies of peripheral lymph node blasts have not yet been completed, but it seems probable that mature immunoblasts taken from this site or from the draining lymph will demonstrate a relative preference for binding to peripheral node HEV. The induction of organ-specific receptors for endothelia may thus serve as a means of directing effector cells back to the areas of the body where they will be most useful – i.e., areas similar to those where the initial antigenic insult occurred.

Gut Intraepithelial Lymphocytes (IEL). Gut IEL are a population of lymphocytes localized between the epithelial cells lining mucosal surfaces, particularly of the small intestine. They are largely nondividing, and roughly 50% contain characteristic cytoplasmic granules. Although their function in vivo remains unknown, as a population they exhibit a significant level of natural killer activity (GUY-GRAND and VASSALLI 1982; ARNAUD-BATTANDIER 1982).

The migratory and HEV-binding properties of gut IEL were of interest because this population is thought to be localized exclusively to mucosal surfaces, and because at least some of them have been shown to derive from dividing precursors in Peyer's patches which migrate, like IgA plasma-cell precursors, to the mesenteric node, the thoracic duct lymph, and then from the blood stream selectively into mucosal sites (GUY-GRAND et al. 1978). In fact, isolated gut IEL, like mesenteric node lymphoblasts, bind with exquisite specificity to Peyer's patch HEV, and poorly if at all to lymph node HEV (M. SCHMITZ, D. NUNEZ, and E.C. BUTCHER, 1986; Table 2).

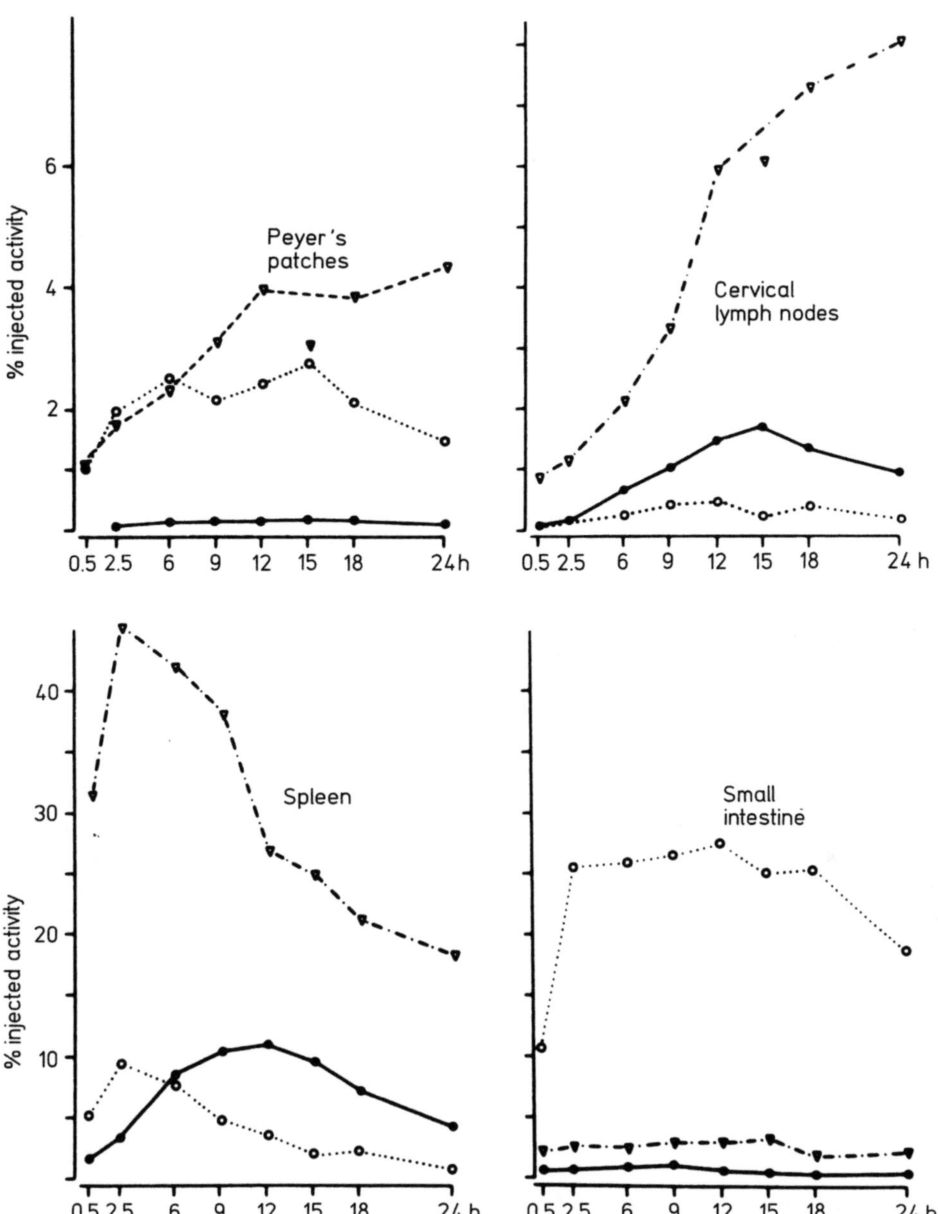

Fig. 5. Localization of IUdR-labeled peripheral node blasts (●), TDL (predominantly gut-derived) blasts (○), and ^{51}Cr-labeled TDL (▽) after injection into syngeneic rats. *Abscissa*, hours after i.v. injection; *ordinate*, percent of injected dose per organ. (From SMITH et al. 1980, with permission)

The HEV-binding specificity of unique mucosal populations argues strongly that lymphocyte-endothelial receptor systems are of fundamental importance in controlling the tissue localization of defined effector cells, and hence in segregating mucosal and nonmucosal immune responses. Furthermore, since gut IEL and gut-derived blasts localize well to intestinal sites distant from Peyer's patches, it seems likely that the same endothelial recognition determinants present on high endothelial postcapillary venules in Peyer's patches (or closely related determinants) may be expressed at some level by other mucosal postcapillary venules, including vessels lined by flat endothelium in the lamina propria. The expression of such a single (mucosal) endothelial specificity for lymphocyte recognition and trafficking could explain the observation that mucosal surfaces appear to be served by a "common mucosal immune system," characterized by dissemination of effector cells to all mucosal tissues regardless of the initial site of antigen stimulation (McDermott and Bienenstock 1979). For instance, induction of mucosal endothelial determinants in the lactating breast could explain the appearance of IgA-secreting plasma cells during lactation and the secretion into milk of IgA antibodies against gut antigens. Similarly, the expression of a nonmucosal endothelial specificity may unify lymph nodes with the skin and perhaps other organs into a common nonmucosal lymphoid system. (Other models of selective traffic to extralymphoid sites must also be considered, however: see Sect. 2.5 below.)

In summary, activated lymphocytes may normally go through two stages of differentiation. The first is a period of suppression of migration during which the lymphocyte remains at the site of stimulation, undergoing proliferation and further differentiation. Possibly as a result of differentiative signals received during this period, at least some lymphoblasts are subsequently induced to reexpress a migratory phenotype, but at this point they express only receptors specific for the endothelial determinant characteristic of the stimulated organ. After leaving the organ in which they arose, these lymphoblasts, presumably antigen-specific effector or regulatory cells, will migrate only to areas of the body that are likely to encounter similar antigenic challenges. The ability to segregate immune effector cells via selective extravasation not only provides an experimental basis for the physiologic distinction between mucosal and nonmucosal immune responses, but may in fact have been a prerequisite for the evolution of unique immune response modalities characteristic of mucosal vs. against nonmucosal sites.

2.1.5 Migratory Properties of Memory Cells

Antigen-reactive lymphoblasts may eventually differentiate either into terminal effector cells, such as plasma cells, or into long-lived memory cells. While it has been shown that some plasmablasts (immunoglobulin-containing immunoblasts) demonstrate organ-specific migration (Guy-Grand et al. 1974; Husband et al. 1977; McWilliams et al. 1977), there is little direct evidence relating to the migratory capacity of terminally differentiated plasma cells. It seems probable that they are a sessile population. On the other hand, it is attractive to hypothesize that *memory* cells may continue to express the same organ- or region-specific migratory characteristics that were initially induced in their lymphoblast precursors. This would insure that memory would be distributed efficiently to regions of the body most likely to require secondary responses to the stimulating antigen.

While it has not been directly tested, two lines of evidence are consistent with the hypothesis that long-lived memory cells may be programed to traffic selectively. First, many small memory cells do recirculate. In fact, long-term B-cell memory appears to be concentrated (on a per cell basis) in lymph (STROBER and DILLEY 1973; reviewed by FORD 1975), suggesting that memory populations may actually recirculate more rapidly than "virgin" lymphocytes. Second, in contrast to the resident small lymphocytes in lymphoid organs, major subpopulations of small lymphocytes obtained from *lymph* exhibit the source-dependent, class-independent organ specificity of migration described above for lymphoblasts and gut IEL (SCOLLAY et al. 1976; CAHILL et al. 1977; CHIN and HAY 1980). Small lymphocytes from lymph draining peripheral lymph nodes in sheep recirculate preferentially through nonmucosal tissues, returning selectively to peripheral lymph. Small lymphocytes in intestinal lymph, on the other hand, return preferentially to intestinal lymph (see Fig. 8). Even purified T cells (from lymph, but not from resident cells in lymph nodes) demonstrate striking selectivity of migration in these studies (CAHILL et al. 1977; REYNOLDS et al. 1982). Furthermore, these source-dependent differences are not observed in the recirculating cells in fetal or neonatal lymph, appearing only after the newborn lamb has experienced antigen and presumably had time to develop memory populations (CAHILL et al. 1980).

2.2 Molecular Basis of Lymphocyte-Endothelial Cell Recognition

2.2.1 A Hypothetical Model of the Mechanism of Selective Lymphocyte-HEV Recognition

The simplest model of lymphocyte-HEV recognition consistent with the cellular data is presented schematically in Fig. 6. According to this model, lymphocyte migration is directed by expression of lymphocyte surface receptors for organ-specific endothelial cell determinants. Two complementary lymphocyte-endothelial cell receptor sets are illustrated, one mediating traffic through the peripheral lymph nodes and possibly other nonmucosal sites, the other through Peyer's patches and appendix. (Additional lymphocyte receptor/endothelial ligand sets regulate lymphocyte traffic through specific extralymphoid sites, e.g., inflamed synovium; see Sect. 2.5.1.) As demonstrated by the nearly absolute specificity shown by certain lymphomas and by gut IEL, some lymphocytes may express receptors only for peripheral node or Peyer's patch HEV determinants, thus directing their migration almost exclusively through peripheral or mucosal sites. The ability of most normal (presumably virgin) B and T cells and T-cell subsets to bind to both lymph node and mucosal HEV may reflect their simultaneous expression of both receptor specificities. The limited organ *preferences* shown

LYMPHOCYTES

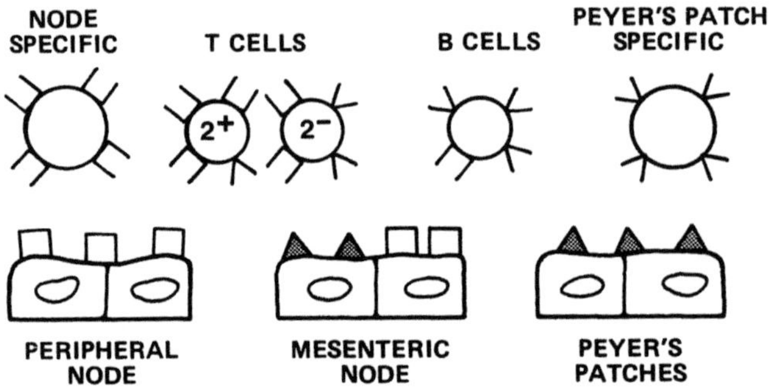

Fig. 6. Model of the hypothetical mechanism of organ-specific lymphocyte migration. (From BUTCHER et al. 1982b, with permission)

by these B- and T-cell populations may be explained by their expression of different proportions of peripheral node and Peyer's patch HEV-specific receptors, as a function of their class. Mesenteric node HEV, which are unique in binding both peripheral node- and Peyer's patch-specific cells, probably express both peripheral node and Peyer's patch HEV determinants (see Sect. 2.4).

It should be mentioned that the lock-and-key recognition proposed is probably an oversimplification of the actual mechanism of lymphocyte-HEV binding. Soluble factors – from lymph (CAREY et al. 1981) or from supernatants of cultured lymph node slides (P. ANDREWS, personal communication; S. WATSON and E.C. BUTCHER, unpublished) – have been identified that can enhance in vitro lymphocyte-HEV interaction, suggesting an additional level of complexity in the regulation of the binding event. Furthermore, LFA-1, a 180/95 mol. wt. glycoprotein complex involved in many leukocyte adhesive events, may also play an accessory role, since monoclonal antibodies against LFA-1 effect partial (30%–50%) inhibition of lymphocyte-HEV binding (HAMANN et al., manuscript in preparation). The functional specificities depicted in the model clearly exist, however, and a minimal requirement for these organ-selective interactions is the existence of separate lymphocyte surface structures mediating recognition of peripheral node and Peyer's patch HEV. Recently, monoclonal antibodies have been produced which provide direct probes for the analysis of such recognition elements.

2.2.2. A gp90 Class of Lymphocyte Surface "Homing Receptors" for Endothelium

A Mouse Lymphocyte Receptor Specific for Lymph Node HEV. We have described a monoclonal antibody, MEL-14, that appears to define mouse lymphocyte surface receptors mediating specific recognition of peripheral lymph node

Table 3. Expression of putative receptor for peripheral node HEV during mouse lymphocyte differentiation

	Staining with MEL-14[a]		
	Approximate % positive	Modal fluorescence[b]	HEV-binding ability (RAR[c] on PN HEV)
Bone marrow pre-B cells	N.D.[d]	60	N.D.
Thymocytes	≥ 90	80	0.05
Peripheral node lymphocytes			
B(IgD$^+$)	>95	600	0.7
T	≥ 80	800	unity
Germinal-center cells	0	0	0.0
Gut IEL	<5	0	0.0

[a] Cell suspensions from BALB/c mice were stained with MEL-14 followed by a highly specific fluoresceinated rabbit anti-rat Ig second stage. Stained cells were analyzed by flow cytofluorimetry. A first-stage rat monoclonal antibody with no known specificity was used as a negative control, defining the background
[b] In arbitrary units above background
[c] See Table 2 notes for definition of RAR
[d] Not determined

HEV (GALLATIN et al. 1983). The antigenic determinant detected by MEL-14 is present on those normal lymphocytes and lymphoma cells which can bind peripheral lymph node HEV, and is absent from lymphoid cells or lymphomas which either do not recognize HEV or bind only to Peyer's patch HEV. Presaturation of lymphocytes with this antibody (but not with other antibodies against abundant cell surface determinants) completely inhibits HEV binding by peripheral node-specific lymphomas, but has no effect on binding of Peyer's patch-specific cells to Peyer's patch HEV. Furthermore, presaturation of normal (dual receptor expressing) lymphocytes with MEL-14 specifically blocks their binding to peripheral node HEV in vitro and selectively inhibits migration into peripheral lymph nodes in vivo, without influencing binding or homing to HEV in Peyer's patches (even though most of the cells binding to Peyer's patch HEV or entering Peyer's patches in vivo are MEL-14$^+$). Although short of formal proof, these studies clearly suggest that the MEL-14 antibody specifically recognizes the lymphocyte receptor for lymph node HEV. Thus there appear to be at least two antigenically as well as functionally distinct HEV receptor specificities, as outlined in the model above – one for peripheral node HEV which is MEL-14$^+$, and one for Peyer's patch HEV which lacks the MEL-14 determinant. The putative lymphocyte receptor for lymph node HEV, precipitated with MEL-14 from lysates of surface iodinated lymphocytes, is an 80–90 kd glycoprotein.

The expression of the MEL-14-defined surface receptor during lymphocyte differentiation correlates well with HEV-binding ability (see Table 3). Pre-B cells in the bone marrow express only low levels of the antigen, as do the majority of thymocytes. A discrete, rare population (1%–3%) of cortical thymocytes, however, stains as brightly as lymph node cells with the antibody (REICHERT et al. 1985). Only a small percentage of thymocytes survive intrathymic selection procedures, and it is interesting to speculate that the MEL-14^{hi} cortical subpopulation may represent or include this exceptional subset of thymocytes destined to emigrate as mature T cells to the periphery. Indeed, when isolated by fluorescence-activated cell sorter, the MEL-14 bright population is highly enriched in functionally mature cytotoxic T-lymphocyte precursors (FINK et al. 1985). Furthermore, the majority of MEL-14 bright cells are phenotypically mature (PNA^{lo}, and expressing either Lyt-2 or L3T4 antigens; REICHERT et al. 1986a).

Expression of the MEL-14 antigen is not, however, entirely limited to functionally and phenotypically mature thymocytes. Most thymocytes express the antigen at very low (non-functional) levels, and a significant subset of MEL-14^{hi} cells are phenotypically immature (PNA^{hi} and expressing both L3T4 and Lyt-2 antigens together, or expressing neither – see REICHERT et al. 1986a). Furthermore, the MEL-14 antigen is expressed at high levels by a major fraction of thymocytes early in thymic ontogeny (day 14–15 of gestation), prior to expression of Lyt-2 or L3T4 differentiation antigens. The significance of the expression of putative homing receptors by these apparently immature populations remains unknown: it may reflect early selection of a lineage of cells destined to populate the periphery. Alternatively, some of these MEL-14^{hi}, Lyt-2$^-$, L3T4$^-$ cells could represent maturing precursors of a different lymphocyte class, for example the "null" phenotype suppressor cells observed in high numbers in neonatal spleen.

Medullary thymocytes are generally thought to represent a mature T-cell population with many features in common with peripheral T cells. They do not participate in the recirculating lymphocyte pool, however, and thus it is not surprising that in immunohistologic studies medullary thymocytes generally exhibit low-to-undetectible staining with MEL-14 (REICHERT et al. 1985). Following treatment with hydrocortisone acetate, however, several interesting alterations in thymocyte homing receptor expression occur. During the phase of cortical involution and death, MEL-14$^+$ thymocytes appear preferentially spared. Within 18 h the medulla, initially MEL-14^{lo} or negative, begins to accumulate increasing numbers of MEL-14$^+$ lymphocytes; the proportion and intensity of staining of MEL-14$^+$ cells increases progressively 2–5 days following hydrocortisone, when a major proportion (35%–40%) of the cells remaining in the thymus – now mostly in the medulla – are intensely MEL-14$^+$ and bind well to HEV. These MEL-14$^+$ medullary thymocytes may represent population of the medulla by descendants of the MEL-14$^+$ cortical cells initially present, and/or they may reflect induction of receptors for HEV on persistent cortisone-resistant medullary cells. (It is unlikely, but also possible, that some of them are derived from mature circulating peripheral T cells.) Clearly, however, the drastic alteration in the predominant phenotype of medullary cells requires reassessment of the classical assumption that cortisone resistant thymocytes and medullary thymocytes are functionally and phenotypically the same. The study of homing receptor expression by thymocyte subsets thus promises to provide significant new insights into intrathymic differentiation processes.

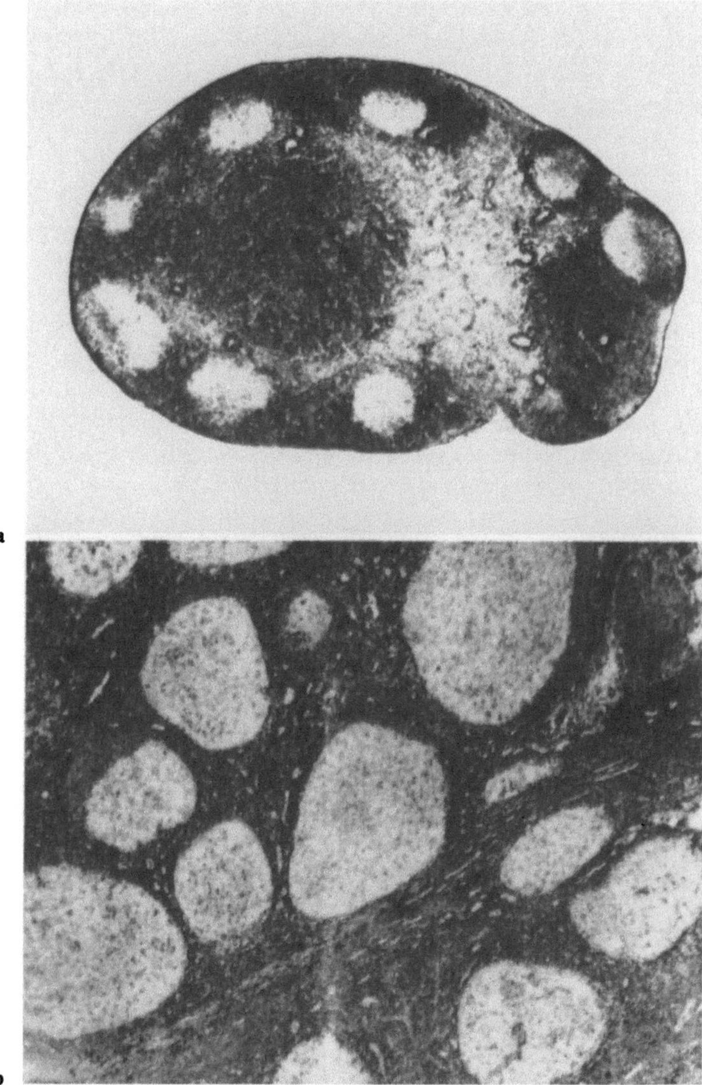

Fig. 7a, b. Immunohistologic (immunoperoxidase) localization of putative homing receptor-expressing lymphocytes. **a** SRBC-immunized mouse lymph node stained with MEL-14. (From REICHERT et al. 1983) **b** Human tonsil stained with Hermes-1. In both mice and humans, germinal centers are largely negative, but surrounding B- and T-cell areas are positive. The medullary region (in **a**) contains only scattered positive cells

Thymic immigrants, identified in lymph nodes or spleen 1–2 h after intrathymic labelling with fluorescein isothiocyanate (FITC), express levels of the MEL-14 antigen comparable with peripheral T cells. Consistent with the model presented in Fig. 6, essentially all IgD$^+$ (circulating) peripheral B lymphocytes, and the majority of peripheral T lymphocytes, are intensely MEL-14$^+$ (see Fig. 7a). Consistent with the preference of T lymphocytes for peripheral node HEV, T lymphocytes bear slightly higher levels of the antigen than do B cells

(Table 3). As might be predicted from the cellular studies reviewed above, the antigen is not expressed on germinal center cells (REICHERT et al. 1983; see Fig. 7a); nor on T-cell clones (DAILEY et al. 1982). In fact, the MEL-14 antigen is rapidly lost from the surface of most mitogen or mixed lymphocyte culture-stimulated T-cell blasts (DAILEY et al. 1983). Splenic marginal zone B cells – a unique population of surface IgM^{hi}, IgD^{lo} B cells found in the marginal zone between the splenic white and red pulps – express only very low levels of the MEL-14 antigen. Low levels of expression of homing receptors (for lymph nodes and other organs) by these marginal zone B cells may explain their sessile nature: they apparently do not circulate to any significant extent, and are not observed in lymph nodes or Peyer's patches (GRAY et al. 1982). Gut intraepithelial lymphocytes, which express functional receptors for Peyer's patch HEV, but not for lymph node HEV (as described above), are MEL-14$^-$ (SCHMITZ et al. 1986). We presume that lymph node-specific normal lymphocyte populations, like lymph node HEV-specific lymphoma cells, will prove to be MEL-14$^+$, but this remains to be determined experimentally.

Human Lymphocyte Homing Receptors. We have recently confirmed that the ability of lymphocytes to interact with HEV is developmentally regulated and organ-specific in humans as it is in mice (JALKANEN and BUTCHER 1985). In addition, we have identified putative human lymphocyte surface receptors for HEV using a novel monoclonal antibody, Hermes-1 (JALKANEN et al. 1986a). Like MEL-14 in the murine system, Hermes-1 immunoprecipitates a roughly 90 kd surface glycoprotein from circulating human lymphocytes. The mouse and human antigens also exhibit a nearly identical, highly acidic isoelectric point (pI about 4.2). The expression of the Hermes-1 antigen is correlated with functional HEV-binding ability by normal lymphocytes and by cell lines. Furthermore, the Hermes-1 antigen has been linked serologically to HEV recognition: highly specific polyclonal antisera produced against the purified Hermes-1 antigen effectively block lymphocyte binding to HEV; thus, antibodies against some epitopes on the Hermes-1 antigen can clearly interfere with the binding event. In a less direct but in some ways more powerful study, we have shown that the affinity-purified Hermes-1 antigen specifically absorbs the functional blocking activity of a whole heterologous rat antihuman lymphocyte serum – thus, of the myriads of antibodies recognizing presumably hundreds of major antigenic epitopes on the surface of human lymphocytes, most antibodies capable of interfering with lymphocyte-HEV binding recognize epitopes present on the Hermes-1-defined molecule (JALKANEN et al. 1986a). Finally, we have found that MEL-14, when included during the in vitro HEV assay, selectively inhibits human lymphocyte binding to lymph node HEV without influencing binding to mucosal (appendix) HEV, indicating a specific if low affinity crossreactivity with functional homing receptors for lymph node HEV. Immunoprecipitation analysis confirms that MEL-14 weakly but selectively crossreacts with the Hermes-1 antigen (JALKANEN et al. 1986c) These findings constitute convincing evidence that the Hermes-1 antigen is functionally important for lymphocyte recognition of lymph node HEV.

Preliminary studies of tryptic cleavage fragments indicate that Hermes-1 recognizes a proximal or central portion of the receptor, distinct from the (presumed recognition) domain bearing the MEL-14 epitope. Consistent with this, Hermes-1 does not directly block lymphocyte binding to HEV. Furthermore, the Hermes-1-defined epitope appears to be present on lymphocyte receptors for mucosal as well as for lymph node HEV: Hermes-1 stains both lymph node-specific cells (B lymphoblastoid line LB25) and mucosal HEV-specific cells (cell line KCA) (see Table 2).

Further support for the presence of the Hermes-1 epitope on multiple receptor classes has come from studies using a polyclonal antiserum, briefly mentioned above, against the Hermes-1 antigen purified from KCA, a human B-lymphoblastoid line that binds preferentially to mucosal HEV. This polyclonal antiserum is highly specific for the Hermes-1-defined molecule; it exhibits cellular staining patterns in immunohistologic sections of lymphoid tissues that are indistinguishable from those of the monoclonal antibody, and precipitates only the same 90-kd molecular specie(s). Most importantly, it effectively blocks lymphocyte binding to lymph note, to appendix and to synovial HEV (JALKANEN et al. 1986c). These results suggest that receptors for different organ-specific HEV determinants, although functionally distinct and independently regulated, are remarkably similar in the human, sharing multiple antigenic epitopes as well as a nearly identical molecular weight and pI. The recent description of rat lymphocyte receptors for Peyer's patch-HEV as glycoproteins of roughly 80 kd (CHIN et al. 1985) supports the general similarities of distinct organ-specific homing receptor classes. These receptor classes may be related by gene duplication and divergence from a common ancestor. Alternatively, the receptors may share a common constant or invariant region, combined by DNA or RNA processing (or by posttranslational modification) with unique variable domains conferring recognition specificity. It remains to be explained why another antibody, A.11, which interferes with rat lymphocyte binding to lymph node HEV, precipitates multiple bands from the rat lymphocyte surface (RASMUSSEN et al. 1985). One possibility is that this antibody, although interacting with lymphocyte receptors for lymph node HEV, recognizes an epitope shared with several other cell surface proteins.

Preliminary studies of the expression of the Hermes-1-defined homing receptor during human lymphocyte differentiation demonstrate significant parallels with, but also interesting differences from, the murine system. Like that MEL-14 antigen in mice, the putative human homing receptor is expressed only at very low levels by most cortical thymocytes, but there is a small subpopulation of cortical cells expressing higher levels which can be identifed in immunohistological staining. Unlike the predominantly MEL-14$^-$ thymic medulla in mice, however, human medullary thymocytes stain intensely with Hermes-1. Since mouse medullary thymocytes can become MEL-14^{hi} following treatment with hydrocortisone, it is possible that the receptor-positivity in the human medulla may reflect the hormonal status of the human donors (young patients undergoing cardiac surgery). Alternatively, it may represent a species difference in the regulation of homing receptors, or may relate to the specificity of Hermes-1 for a common determinant present on multiple receptors. For example, it is possible that resident medullary cells in both species lack the receptor for lymph node HEV, as defined in the mouse by MEL-14, but express high levels of mucosal HEV receptors or of another related receptor, perhaps involved in "homing" to medullary region of the thymus.

Mature, circulating peripheral B and T cells express high levels of the antigen but, as in the mouse, activated B cells in germinal centers are largely negative for the putative receptor (see Fig. 7b). In keeping with their capacity to bind to HEV, human T-cell clones, unlike their MEL-14$^-$ murine counterparts, express high levels of Hermes-1-defined homing receptors, suggesting that homing receptors may be regulated differentially in mice and humans, at least within the T cell lineage (JALKANEN et al. 1986a). Further studies of homing receptor expression should define additional parallels and differences between species, and may shed light on the nature and rate of evolutionary changes in the regulation of homing receptors, and of cell interaction receptor systems in general.

2.3 Involvement of Homing Receptor-Related Molecules in the Extravasation of Other Leukocytes

The expression of the antigens defined by MEL-14 and Hermes-1 is not limited to lymphoid cells. Both MEL-14 and Hermes-1 stain mature granulocytes and monocytes intensely, and immunoprecipitate a neutrophil surface glycoprotein of about 100 kd, slightly larger than the lymphocyte homing receptors (LEWINSOHN et al. 1986). The antigenic crossreactivity and biochemical similarity of the neutrophil and lymphocyte antigens suggest that they may be related functionally, and in fact we have found that MEL-14 blocks neutrophil interaction with endothelial cells in vitro (in the frozen section assay of binding to lymph node endothelium), and inhibits neutrophil extravasation from the blood into peripheral sites of inflammation – for example, in the skin. Since neutrophils do not normally interact with HEV, it is clear that either a) the neutrophil protein is related to but is not functionally identical to the putative lymphocyte receptor for HEV; or b) if these molecules are identical in their binding domains, accessory mechanisms must be required for binding to HEV or other endothelial cells (in addition to endothelial cell recognition via the Hermes-1/MEL-14-defined gp90/100 receptors). Such accessory mechanisms could be constitutively expressed by lymphocytes, but might require activation or triggering by inflammatory factors to allow neutrophils to interact with endothelium and then extravasate in sites of acute inflammatory insult (BUTCHER et al. 1986).

MEL-14 and Hermes-1 thus appear to define functionally important, developmentally regulated 90/100 kd receptors necessary not only for the organ-specific trafficking of lymphocytes, but also for the accumulation of neutrophils and other leukocytes in sites of acute and chronic inflammation. By controlling the extravasation of lymphocytes and other leukocytes in particular lymphoid tissues or inflammatory sites, these receptors for endothelium thus play a pivotal role in determining the extent and nature of local immune/inflammatory reactions.

2.4 Nature and Regulation of HEV Determinants Mediating Lymphocyte Recognition

Little is known about the endothelial cell surface determinants recognized by migrating lymphocytes. Recently reported experiments, however, suggest that a carbohydrate component may play an important role in lymphocyte-HEV interactions: ROSEN et al. (1985) have shown that neuraminidase treatment of frozen sections completely abolishes lymphocyte binding to lymph node HEV, but has no influence on binding to mucosal HEV (in Peyer's patches), suggesting that sialic acid per se (or perhaps charge effects) may be important in specific recognition and binding to lymph node HEV determinants. STOOLMAN et al. (1984) have shown that mannose-6-phosphate is a potent inhibitor of lymphocyte-HEV-binding in murine species, yielding 50% inhibition in the millimolar range. The sugar is thought to inhibit a lymphocyte surface molecule, since pretreatment of lymphocytes with low levels of a mannose-6-phosphate rich phosphomannan core protein (from *Hansenula holstii*) also effectively inhibits binding. KIEDA and MONSIGNY (1983) have reported that beta-galactosyl-containing neoglycoproteins, such as beta-lactosyl-bovine serum albumin, are also effective inhibitors of lymphocyte-HEV binding in the mouse. These studies have been interpreted as suggesting a lectin-carbohydrate-based recognition system for lymphocyte-HEV binding.

In direct support of this proposal, Rosen and coworkers have recently shown a) that the effect of phosphomannan is organ-specific, inhibiting binding of lymphocytes to lymph node but not to Peyer's patch HEV; b) that phosphomannan-conjugated beads bind specifically to lymph node HEV-binding (MEL-14$^+$) mouse lymphocytes and cell lines, and fail to bind to HEV non-binding cells; and c) that binding of beads is blocked by preincubation of lymphocytes with MEL-14 (YEDNOCK et al. 1986). These results clearly suggest that the MEL-14-defined lymphocyte homing receptor can function as a mammalian lectin with specificity for mannose-6-phosphate. It seems likely therefore that homing receptors may interact with organ-specific carbohydrate determinants on endothelium. The interaction of homing receptors with mannose-6-phosphate is of low affinity, however, and thus it remains formally possible that the actual endothelial cell ligand, although mimicked by mannose-6-phosphate, is an unrelated molecular structure.

ANDREWS et al. (1980, 1982) have taken another approach to studying the role of HEV in lymphocyte-HEV interactions. They observed that murine HEV cells actively take up $^{35}SO_4$ and incorporate it into a glycosylated macromolecule (possibly a glycolipid) that is released or secreted in noncovalent association with a pronase-sensitive component (ANDREWS et al. 1983). Preliminary functional studies suggest that this complex may enhance lymphocyte-HEV binding in vitro (P. ANDREWS, personal communication), and may induce lymphocytes to accumulate at intradermal sites of injection (ANDREWS et al. 1980). Examination of the organ-specificity of this enhancing effect may help determine whether this putative HEV-produced factor is directly involved in lymphocyte recognition of HEV, or instead may nonspecifically influence or facilitate lymphocyte binding triggered by separate, specific recognition mechanisms.

While some cell-cell adhesion events may be mediated by like-like interactions at the molecular level, this is unlikely in the case of lymphocyte-HEV recognition since mouse lymph node HEV do not express detectable levels of the MEL-14 antigen, and human HEV cells are Hermes-1⁻.

Regardless of the specific nature of the endothelial cell mechanisms mediating lymphocyte-HEV interaction, it is becoming increasingly clear that their expression is controlled by factors reflecting local requirements for lymphocyte traffic. Significant increases in the quantity of HEV occur in antigen-stimulated lymph nodes (ANDERSON et al. 1975). Morphologically well-defined HEV mediating lymphocyte traffic from the blood can be induced in sites of chronic inflammation throughout the body in many species, including humans. Furthermore, in lymph nodes surgically deprived of afferent lymphatics, the high endothelial lining of HEV progressively disappears over several weeks, reverting eventually to a flat endothelium incapable of supporting lymphocyte extravasation from the blood (HENDRIKS and ESTERMANS 1983). Stimulation of isolated lymph nodes with xenogeneic red blood cells resulted in reemergence of morphologically and functionally identifiable HEV.

It seems likely that specific cytokines may be important in controlling high endothelial cell differentiation. We have recently described a monoclonal antibody, MECA-325, defining a mouse endothelial differentiation antigen that is specifically expressed on HEV in organized lymphoid tissues, and that appears to be induced in association with high endothelial differentiation and increased lymphocyte traffic in extralymphoid sites of chronic inflammation (DUIVESTIJN et al. 1986). This HEV-specific differentiation antigen can be induced in endothelium derived from mouse lung by interferon-gamma. This finding suggests that the high endothelium is an inducible phenotype of endothelial cells, rather than a developmentally distinct endothelial cell class or lineage. It also suggests that γ-interferon alone or (more likely) in combination with other cytokines may play an important role in inducing and maintaining HEV, and thus in regulating lymphocyte traffic both in lymphoid organs and in extralymphoid inflammatory sites. Interleukin-1 may also be important in regulating endothelial cell differentiation in this context, since this interleukin has been shown to enhance the binding of lymphocytes to cultured human endothelium (BEVILACQUA et al. 1985; CAVENDER et al. 1986). Glucocorticoids, which decrease lymphocyte localization to lymph nodes and increase distribution to the bone marrow, also appear to influence lymphocyte traffic by acting directly or indirectly on HEV: in mice receiving continuous low doses of glucocorticoids, HEV lose their ability to bind lymphocytes (R. DAYNES, personal communication).

Cellular or humoral factors in local tissues or draining lymphatics may also be able to influence the *specificity* of lymphocyte homing determinants expressed by endothelial cells: The mesentric node, which receives its afferent lymphatic supply from the intestines, is unique among lymph nodes examined, in that it contains HEV capable of binding Peyer's patch-specific as well as lymph node-specific lymphoma cells. Thus, intestinal lymph may contain cells or factors inducing mucosal-type endothelial determinants on some HEV in the mesenteric node. (Individual HEV or segments of HEV can be identified in the mesenteric node which bind exclusively lymph node-specific or Peyer's patch-specific lymphocytes, and this specificity is maintained in serial sections of the same segment. Thus, the expression of particular endothelial cell determinants may be either clonally determined, or under micro-regional control.) This finding, along with the observation that fetal gut explants grown underneath adult kidney capsules attract the same lymphocyte populations that home to adult gut (HALSTEAD and HALL 1972; PARROT and FERGUSON 1974; GUY-GRAND et al. 1974), suggests that the specificity of endothelial cells may be determined by local microenvironmental influences or factors, probably entirely independent of the presence of lymphocytes or antigen.

2.5 Lymphocyte Migration Through Nonlymphoid Tissues

Although lymphocytes recirculate most often through the organized lymphoid tissues, there is a small but significant lymphocyte traffic through other tissues as well (SMITH et al. 1970; RANNIE and DONALD 1977; STREILEIN 1978). Two of the most important organs in this respect are the gut (especially the lamina propria of the small and large intestine) and the skin. Although relatively few lymphocytes leave the blood at these sites (RANNIE and DONALD 1977) – at least when expressed per gram of tissue or per unit of blood flow – lymphocyte traffic through the gut and skin exhibits significant parallels to lymphocyte migration through their associated lymphoid organs, the Peyer's patches and the peripheral lymph nodes. This similarity is most notable in studies of the migration of mature immunoblast populations. Those blasts (e.g., from stimulated peripheral lymph nodes) which localize preferentially in peripheral nodes as opposed to Peyer's patches also migrate selectively to sites of inflammation in the skin (HALL et al. 1979; ROSE et al. 1978). Conversely, mesenteric node (gut-derived) blasts localize in large numbers in the gut wall but avoid the skin (ROSE et al. 1978). These observations further support the concept, introduced in Sect. 2.1.4, that endothelial cells in nonlymphoid tissues may express the same determinants for lymphocyte recognition that are present on HEV, but presumably in lower concentrations.

There are, however, qualitative as well as quantitative differences in the migration of lymphocyte subsets to nonlymphoid vs. lymphoid tissues. For instance, selectively migrating immunoblasts generally localize much better than small lymphocyte populations in nonlymphoid inflammatory sites, even though both populations may localize equally well (or the small lymphocytes may even localize more efficiently) in the appropriate lymphoid organs (GOWANS and KNIGHT 1964; SMITH et al. 1980; OTTAWAY and PARROTT 1979; HALL et al. 1972; ASHERSON and ALLWOOD 1972). For instance, as illustrated in Fig. 5, thoracic duct (gut-derived, mature, migrating) immunoblasts localize as well as whole thoracic duct lymphocytes (TDL, predominantly small lymphocytes) in Peyer's patches early after i.v. injection, but migrate much more efficiently than whole TDL to the gut wall. Similar results have been reported for peripheral node blasts, which localize better than small lymphocyte populations in sites of inflammation in the skin (OTTAWAY and PARROTT 1979). Although these observations might imply the existence of additional mechanisms for controlling the specificity of lymphocyte migration in nonlymphoid tissues, in fact they might also be explained (at least theoretically) on the basis of differences in the density of expression of the same lymphocyte and endothelial recognition determinants mediating lymphocyte interaction with HEV. If multiple lymphocyte receptor-HEV determinant interactions are required for a successful binding event, and if endothelial determinants are limiting in nonlymphoid tissues, then only those lymphocytes expressing increased numbers of complementary receptors would be expected to adhere to endothelium in extralymphoid sites. Differences in the surface density of lymphocyte homing receptors would make less difference in binding to HEV where endothelial determinants are in relative abundance. In the context of this model we might presume, for instance, that immunoblasts expressing a *single* receptor type might bear a higher surface density of that receptor than would "virgin" cells of dual HEV specificity. Such an increased density of surface receptors on organ-specific lymphocyte populations could support their increased traffic through the (non-HEV) endothelium of extralymphoid tissues.

Such a model would predict that not just immunoblasts, but any lymphocyte expressing receptors exclusively for peripheral node- or Peyer's patch-HEV determinants, might bear a higher surface density of specific receptors and thus should migrate more frequently through appropriate nonlymphoid tissues than would "virgin" lymphocytes with dual receptor expression. One might expect, therefore, that the selectively recirculating small lymphocyte populations observed in peripheral lymphatics in the sheep might exhibit even greater specificity of recirculation through skin than through the peripheral lymph nodes (since admixed nonspecifically recirculating cells should be less well represented among cells trafficking through the skin). This prediction has been verified by the results of CHIN and HAY (1980), who studied the differential recirculation of cells from intestinal vs. peripheral (afferent) lymph through sites of inflammation in the skin, through peripheral lymph nodes, and through the intestine. As shown in Fig. 8a, lymphocytes from the afferent lymph draining a skin granuloma (labeled in vitro with [111]Ind and injected i.v. into the same animal) return much better to peripheral efferent lymphatics (draining the prefemoral lymph node) than to intestinal lymphatics, but return in even greater specific activity to afferent lymph draining the chronic inflammatory site. In contrast, lymphocytes from intestinal lymphatics circulate preferentially through the intestine, migrating poorly through both the prefemoral lymph node and the peripheral granuloma (Fig. 8b). These results suggest that selectively migrating effector and memory populations may pass much more frequently than their virgin precursors through nonlymphoid tissues.

Two attractive alternative hypotheses that might explain the differential migration of lymphoblasts to inflammatory sites must also be considered. First, such selectively migrating lymphocyte populations may interact with extralymphoid endothelium via the same endothelial recognition systems defined above, but the use of these receptors might be controlled or triggered by accessory signaling systems specifically present in inflammatory sites (as proposed for neutrophils, above). Thus, circulating effector or memory cells could recognize mucosal vs. nonmucosal vascular beds via organ-specific receptors for endothe-

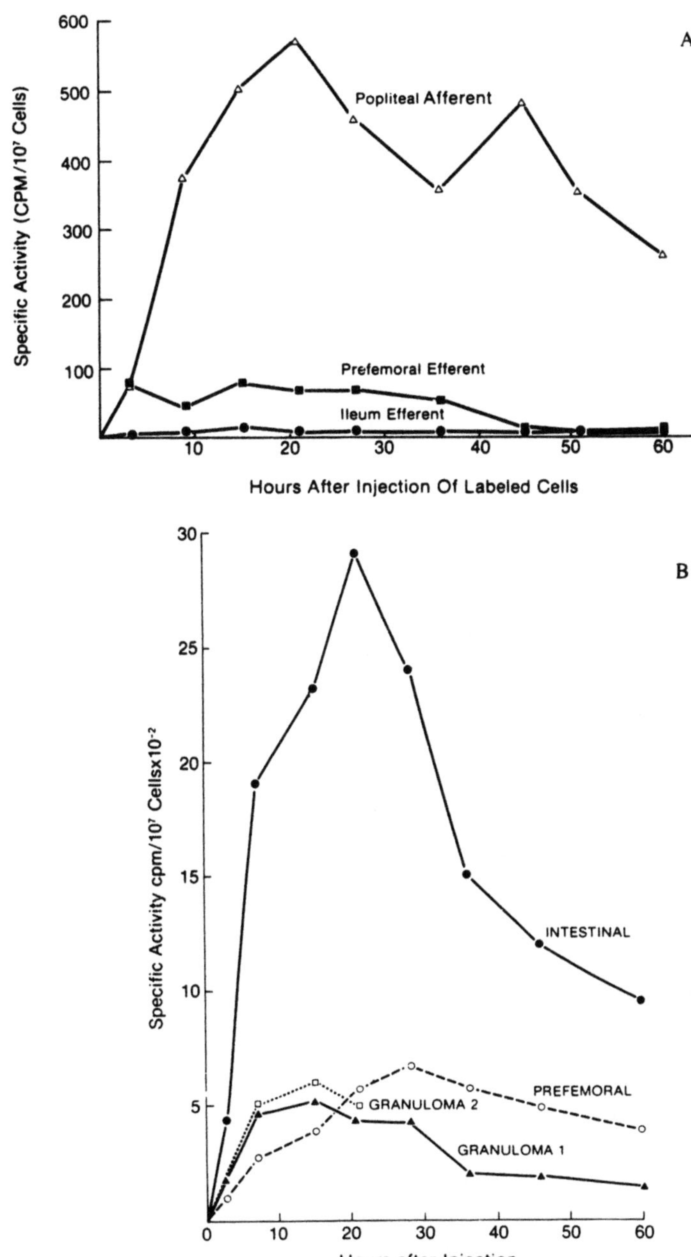

Fig. 8A. Specific activity profiles in the afferent lymph draining a granuloma (△–△), efferent prefemoral lymph (■–■) and the efferent lymph draining the ileum (●–●), following i.v. injection of 7.2×10^7 afferent lymph cells labeled with 1.4×10^7 cpm ^{111}In. (From CHIN and HAY 1980, with permission). **B** Specific activity profiles in efferent intestinal lymph (●–●), efferent prefemoral lymph (○–○), and afferent lymph draining a granuloma in the right hind leg (▲–▲) and the left hind leg (□–□), following i.v. injection of 4.0×10^8 efferent intestinal lymphocytes labeled with 3×10^7 cpm ^{111}In. (From CHIN and HAY 1980, in permission)

lium, and could be signaled to utilize these receptors (or to activate other cellular systems required for adherence to and/or migration through the endothelium) by soluble or endothelial surface-associated factors present at sites of tissue inflammation. (Similarly, lymphocytes trafficking through HEV in lymph nodes and Peyer's patches might require a distinct HEV-derived or organized lymphoid tissue-associated signal leading to receptor utilization and extravasation.) Another possibility is that migration through some or all extralymphoid sites is controlled by additional lymphocyte-endothelial receptor/ligand systems, distinct from those operating in lymph nodes and mucosal lymphoid organs. For example, the homing of IEL and gut-derived blasts to the gut wall could reflect expression of a lamina propria-specific homing receptor. This model would predict the existence of a larger family of lymphocyte surface receptor/endothelial cell ligand sets, regulating lymphocyte traffic in a specific manner through many different tissues of the body.

2.5.1 A Distinct Endothelial Cell Recognition System Controlling Lymphocyte Traffic to Inflamed Synovium

In support of the role of additional endothelial cell ligands in controlling lymphocyte traffic to inflamed extralymphoid sites, we have recently demonstrated that lymphocytes bind to HEV in inflamed synovial tissue via a recognition system that is functionally distinct from homing receptors for lymph node and mucosal HEV (JALKANEN et al. 1986b). HEV appear in association with chronic inflammatory infiltrates in the synovial lining of joints in patients with rheumatoid arthritis. These HEV bind peripheral blood lymphocytes (PBL) efficiently, but fail to interact with either the lymph node-specific cell line LB25, or the mucosal HEV-specific cell KCA (see Table 2). In addition, PBL binding to synovial HEV, unlike adherence to lymph node HEV, is not inhibited by MEL-14. It is blocked, however, by the polyclonal anti-Hermes-1 antigen serum described in Sect. 2.2.2, indicating involvement of the gp 90 class of putative homing receptors (JALKANEN et al. 1986c). While we do not as yet known whether the specificity observed is synovial tissue-specific, or instead represents a mechanism for homing to inflamed tissues in general, the existence of three functionally independent lymphocyte-endothelial receptor-ligand sets clearly suggests that there may be a larger family of recognition systems regulating lymphocyte traffic into many different organs and tissues. Organ-specific lymphocyte homing mechanisms may be important not only in enhancing the efficiency of immune responses in particular types of tissues, but also in decreasing opportunities for autoimmune crossreactions by excluding, for example, effector cells arising in response to skin pathogens from entering relatively privileged sites such as joints.

3 Mechanisms Controlling Lymphocyte Migration to the Spleen

The spleen is a relatively complex organ composed of two physically associated but functionally distinct compartments – the organized lymphoid tissue of the white pulp, and the surrounding red pulp, which has a variety of functions including hematopoiesis and the scavenging of old red blood cells. Little is known of the cellular mechanisms involved in lymphocyte migration to the splenic white pulp. The spleen contains no histologic equivalent of the HEV seen in other organized lymphoid organs, and the cells responsible for trapping migrating lymphocytes have not been identified. Although most lymphocytes capable of localizing in lymph nodes or Peyer's patches are also capable of lodging in the spleen (in either the red or white pulp), two lines of evidence suggest that the splenic homing mechanism may be distinct from that of peripheral node and Peyer's patch HEV. First, the pattern of short-term localization of the major "virgin" lymphocyte populations to the spleen is quite distinct from their pattern of migration to Peyer's patches or peripheral lymph nodes (STEVENS et al. 1982; KRAAL et al. 1983; SPRENT 1973). B cells localize much better than T cells in the spleen 2 h after i.v. injection. In this respect, the spleen is not unlike the Peyer's patches – although B cells localize even better, relative to T cells, in the spleen than in the Peyer's patches (see Fig. 3; STEVENS et al. 1982). Splenic localization of migrating T-cell subsets, however, more closely reflects their homing to peripheral lymph nodes than to the Peyer's patches – Lyt-2$^-$ T cells migrate about as well as Lyt-2$^+$ cells to both the spleen and to the peripheral nodes, but significantly better (1.5–2 times) to the Peyer's patches (KRAAL et al. 1983). Thus, the mechanisms mediating the specificity of migration to the spleen must be different from those of the known HEV determinants. Second, the mechanism of localization of lymphocytes to the splenic white pulp can be distinguished from those mediating interactions with lymph node HEV on the basis of sensitivity to protease treatment. Lymphocyte interactions with lymph node HEV can be almost completely abolished by gentle treatment of lymphocytes with trypsin (WOODRUFF et al. 1977). The same levels of trypsinization have little or no effect on the entry of lymphocytes into the splenic white pulp (FORD et al. 1978). (Interestingly, when trypsinization is performed in the presence of Ca^{++}, lymphocyte binding to mucosal HEV is also spared – S. ROSEN and T. YEDNOCK, personal communication.) It will be interesting to determine if splenic homing receptors are also members of the gp90 receptor class.

Regardless of the nature of the mechanisms controlling lymphocyte entry into the spleen, it seems clear that they are important in controlling the splenic representation of lymphocyte subsets. The increased localization of B (vs. T) cells is reflected in a predominance of B cells in the spleen compared with in the lymph nodes (STEVENS et al. 1982), and the similarity of the relative localization of the T-cell subsets in the spleen and the peripheral lymph nodes is reflected in the similar proportions of Lyt-2$^-$ and Lyt-2$^+$ T cells in these organs, in contrast to the increased representation of Lyt-2$^-$ cells in the Peyer's patches (KRAAL et al. 1983).

4 Role of Antigen in the Migration and Tissue Distribution of Lymphocytes

Antigenic stimulation influences the localization of lymphocytes in a variety of ways. One of the earliest and most striking effects of antigen is to increase blood flow through the stimulated tissues, resulting in a significant increase in the rate of delivery of circulating lymphocytes [reviewed by HAY et al. (1980)]. Hyperemia, which begins within hours after stimulation, may be mediated by release of soluble factors – probably prostaglandins or related products of arachidonic acid metabolism (HAY et al. 1980). The blood flow to a local granulomatous reaction in skin or to a stimulated lymph node may reach 10–25 times normal levels (HAY et al. 1980), resulting in a parallel increase in lymphocyte traffic, and a dramatic increase in the total number of lymphocytes in the tissue.

Decreases in the rate of lymphocyte transit *through* lymph nodes can also greatly influence lymphocyte accumulation following antigen stimulation; in fact, certain antigens are able to cause a near shutdown in lymphocyte exit from stimulated lymph nodes (CAHILL et al. 1976; TRNKA and CAHILL 1980). This control of lymphocyte exit may be mediated, at least in part, by leukocyte interferon and/or prostaglandin E2. For instance, interferon reduces the thoracic duct lymphocyte output (KORNGOLD et al. 1983) and causes lymph node enlargement in mice (GRESSER et al. 1981). Prostaglandin E2 has also been shown to inhibit lymphocyte exit from lymph nodes (HOPKINS et al. 1981). Other effector substances may also be involved, perhaps acting through a common intracellular mechanism, involving increased lymphocyte transit following elevations in intracellular cyclic GMP and inhibition of lymphocyte transit following elevations of cyclic AMP (MOORE and LACHMANN 1982). Complement activation within lymph nodes also initiates decreased cell exit (MCCONNELL and HOPKINS 1981). The combined effect of these mechanisms increasing lymphocyte influx and inhibiting lymphocyte exit can be truly dramatic. A lymph node may reach 15 times its normal size within 5–7 days after antigenic stimulation (HAY et al. 1980). *Hyperemia and the control of lymphocyte transit are thus important mechanisms that greatly enhance the number of lymphocytes available for immune responses at sites of antigenic insult.*

In addition to these nonspecific alterations which may affect all circulating lymphocytes, antigenic stimulation also exerts a profound selective influence on the local representation of *antigen-specific* lymphocytes. This influence does not appear to be mediated by selective migration of antigen-specific lymphocytes from the blood, but seems to operate as follows: first, by "trapping" or inhibiting the exit of circulating lymphocytes from areas where they encounter specific antigen (SPRENT 1980; ROWLEY et al. 1972), although this point is controversial (MCCULLAGH 1980); and second, by inducing the local proliferation and expansion of antigen-specific lymphocyte populations. Antigen-induced alterations in the distribution of antigen-specific lymphocytes can affect both B and T cells, can occur in both lymphoid and nonlymphoid sites of antigen deposition, and probably involve cellular mechanisms of antigen presentation resulting in induction of nonmigratory characteristics (for instance, suppression of receptors for HEV) and frequently blastogenesis. The following examples serve to illustrate these points.

After primary immunization of a lymph node, the frequency of antigen-binding B cells rapidly increases. These antigen-binding cells are easily studied by day 5–7 after immunization with keyhole limpet hemacyanin (KRAAL et al.

1986) or sheep red blood cells (SRBC) (KRAAL et al. 1982), at which time most of them have the unique (IgD⁻, peanut agglutinin-binding) phenotype of germinal-center B cells. As is characteristic of germinal-center cells, they express low-to-undetectable levels of surface receptor for peripheral node HEV (as determined by immunofluorescence staining with the MEL-14 antibody, described above). The antigen-binding lymphocytes are predominantly large, presumably dividing cells. Whether the initial increase in antigen-binding cells is due to trapping of passing antigen-specific "virgin" cells or simply to proliferation of a subset of these cells has not been directly examined in this system. It is clear, however, that antigen is responsible for significantly altering the migratory cycle of many antigen-specific B cells that remain to become incorporated in germinal centers.

A parallel sequence of events affects antigen-specific helper T cells migrating through antigen-stimulated lymphoid organs (reviewed by SPRENT 1980), and in this case actual trapping of specific T helper cells has been demonstrated. For 1–2 days after i.v. administration of SRBC, T cell help for the adoptive anti-SRBC humoral response is specifically depleted from the thoracic duct lymph of the stimulated animal, but is increased in the spleen (SPRENT 1980). After 3 days, antigen-specific T cell help reappears in lymph and soon is present in much higher levels than before immunization. These observations suggest that circulating antigen-specific T cells are selectively arrested in the spleen (a major site of antigen deposition after i.v. injection), where they are induced to divide and differentiate, eventually returning to the circulating lymphocyte pool. Lymphocyte trapping in the immunized spleen, at least in this experimental setting, requires H-2 compatibility between the circulating T cells and radioresistant cells in the recipient's spleen (SPRENT 1980). Thus, the arrest of specific lymphocytes by antigen almost certainly involves H-2-restricted interaction with antigen-presenting cells, presumably resulting in induction of nonmigratory phenotypic characteristics as well as proliferation.

The effect of antigen on the localization of mature antigen-specific plasma cell precursors is well illustrated in Fig. 9. In this experiment, HUSBAND (1982) surgically created two entirely independent loops of small intestine (Thiry-Vella loops) in rats that had been primed i.p. with purified tetanus toxoid and ovalbumin 10 days earlier. Four days later, the thoracic duct was cannulated and the intestinal loops were challenged – toxoid was injected into the lumen of the distal loop and ovalbumin into that of the proximal loop. TDL collected from day 0 to day 3 after immunization were discarded – thus, neither the proximal nor the distal loops received significant numbers of plasma cell precursors during this time. TDL collected between days 3 and 4 after challenge were returned i.v. into each rat, and the density of specific antitoxin-containing cells (ATCC, defined by cytoplasmic binding of tetanus toxoid) and of antiovalbumin-containing cells (AOCC) was determined in histologic sections of the proximal and distal loops. The data (Fig. 9) demonstrate that the initial migration of plasmablasts into the lamina propria is antigen-independent – there is no difference in the frequency of ATCC and AOCC in the two loops at 3 or 6 h after injection. Thereafter, however, there is a striking increase in the density of ATCC in the tetanus-toxoid-challenged loop, as opposed to an

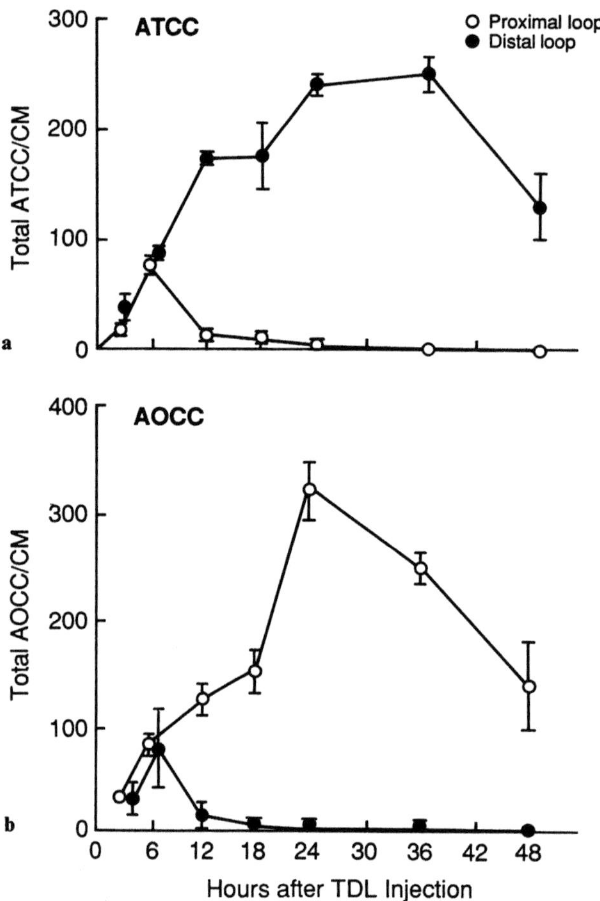

Fig. 9a. Density of antitoxin-containing cells (*ATCC*) or **b.** antiovalbumin-containing cells (*AOCC*) in proximal loop (o) and distal loop (●) of primed rats at various times after an injection of TDL; see text for explanation. (From HUSBAND 1982, with permission)

actual decline in AOCC. Conversely, there was an increase in AOCC and decrease in ATCC in the ovalbumin-challenged loop. Since the thoracic duct was continuously drained during this experiment, these results cannot be explained by recirculation of antibody-containing cells. The increase in specific antibody-containing cells in the appropriate challenged loop must be attributed to a combination of antigen-induced cell trapping and local proliferation; in fact, a significant proportion of plasma cells became labeled when ^{3}H-thymidine was administered during the period of rapid increase in specific antibody-containing cell (ACC) density illustrated in Fig. 9. The decline in the density of ACC in the inappropriately challenged loop may illustrate either a failure of cell trapping (an increased rate of transit into lymph), cell death, or both in the absence of appropriate antigenic stimulation.

In conclusion, localized antigen acts through as yet ill-defined cellular and humoral mechanisms to increase blood flow and hence lymphocyte traffic through sites of inflammation, to slow the transit of cells back to lymph from the stimulated area, and to arrest the migration and induce the local proliferation and differentiation of antigen-specific B and T cells. One antigen-induced differentiative event may involve reprogramming of the specificity of lymphocyte migration, as suggested above, so that antigen-experienced memory and effector populations would return to appropriate tissues after leaving the site of stimulation.

5 Summary

Two principal mechanisms controlling the tissue distribution of lymphocytes have been described: a) lymphocyte-endothelial cell recognition, which controls lymphocyte extravasation and the flow of lymphocyte subsets through mucosal, peripheral lymph node, synovial, and probably other tissues in a highly selective manner; and b) as yet poorly defined cellular mechanisms of antigen processing, which lead to local increases in lymphocyte traffic, and which selectively arrest the exit into lymph and induce the local proliferation of antigen specific lymphocyte populations. The selective expression of 90 kd lymphocyte surface receptors for tissue- or organ-specific endothelial determinants regulates the migratory characteristics of lymphocytes at various stages of differentiation, and appears to be a major element determining the representation of lymphocyte subsets in particular sites. The major populations of small, mature, presumably "virgin" lymphocytes appear to express all three identified homing receptor classes (for lymph node, mucosal, and synovial endothelium), but exhibit certain endothelial recognition (and hence organ distribution) preferences that are a function of lymphocyte class. Antigen acts through cellular mechanisms to alter the distribution of *antigen-specific* B and T cells, probably by inducing a transient nonmigratory phenotype in circulating antigen-specific cells, and certainly by causing their local proliferation. The evidence suggests that proliferating antigen-specific cells may be induced by local microenvironmental influences to express surface receptors specific for the type of endothelial determinant associated with the stimulated organ. Expression of a single specificity of homing receptors thus appears to direct the migration and restrict the tissue distribution of many effector and memory cells to regions of the body most likely to reencounter the stimulating antigen. This restricted migration of effector and memory cells is probably important a) in enhancing the efficiency of immune responses in particular tissues or organs; b) in unifying immunity in related tissues – for example, the mucosal surfaces; and c) in decreasing opportunities for autoimmune crossreactions by excluding irrelevant effector cells from relatively privileged sites such as joints.

Acknowledgments. I thank all the people who have shared my interest in and work on the mechanisms of lymphocyte migration, especially S.T. Jalkanen, R. Bargatze, G. Kraal, A. Duijvestijn, R. Reichert, W.M. Gallatin, and I.L. Weissman. Special thanks are extended to D. Butcher and C. Butnut for continuing support and encouragement. The original work described has been supported by NIH grants AI 19957, CA 34709, and AI 09072, and by a grant from the Veterans Administration. Portions of this manuscript were reprinted with permission from BUTCHER (1983).

References

Anderson ND, Anderson AO, Wyllie RG (1975) Microvascular changes in lymph nodes draining skin allografts. Am J Pathol 81:131–160

Andrews P, Ford WL, Stoddart RW (1980) Metabolic studies of high-walled endothelium of post-capillary venules in rat lymph nodes. In: Blood cells and vessel walls. Excerpta Medica, Amsterdam, pp 211–230 (Ciba Foundation symposium no 71)

Andrews P, Milsom DW, Ford WL (1982) Migration of lymphocytes across specialized vascular endothelium. V. Production of a sulphated macromolecule by high endothelial cells in lymph nodes. J Cell Sci 57:277–292

Andrews P, Milson DW, Stoddart RW (1983) Glycoconjugates from high endothelial cells. I. Partial characterization of a sulphated glycoconjugate from the high endothelial cells of rat lymph nodes. J Cell Sci 59:231–244

Arnaud-Battandier F (1982) Immunologic characteristics of isolated gut mucosal lymphoid cells. In: Strober W, Hanson LA, Sell KW (eds) Recent advances in mucosal immunity. Raven, New York, p 289

Asherson GL, Allwood GG (1972) Inflammatory lymphoid cells. Cells in immunized lymph nodes that move to sites of inflammation. Immunology 22:493

Bevilacqua MP, Pober JS, Wheeler ME, Cotran RS, Gimbrone MA (1985) Interleukin-1 acts on cultured human vascular endothelium to increase the adhesion of polymorphonuclear leukocytes, monocytes, and related leukocyte cell lines. J Clin Invest 76:2003–2011

Butcher EC (1983) The control of lymphocyte migration and tissue distribution. In: Daynes RA, Spikes JD (eds) Experimental and clinical photoimmunology, vol 1, chap. 12. CRC, Boca Raton

Butcher EC, Weissman IL (1979) Cellular, genetic and evolutionary aspects of lymphocyte interactions with high endothelial venules. In: Blood cells and vessel walls: functional interactions. Excerpta Medica, Amsterdam, pp 265–286 (Ciba Foundation symposium no 71)

Butcher EC, Scollay RG, Weissman IL (1979) Lymphocyte adherence to high endothelial venules: characterization of a modified in vitro assay, and examination of the binding of syngeneic and allogeneic lymphocyte populations. J Immunol 123:1996

Butcher EC, Scollay RG, Weissman IL (1980) Organ specificity of lymphocyte migration: mediation by highly selective lymphocyte interactions with organ-specific determinants on high endothelial venules. Eur J Immunol 10:556

Butcher EC, Kraal G, Stevens SK, Weissman IL (1982a) Selective migration of murine lymphocyte and lymphoblast populations and the role of endothelial cell recognition. Adv Exp Med Biol 149:199–206

Butcher EC, Kraal G, Stevens SK, Weissman IL (1982b) A recognition function of endothelial cells: directing lymphocyte traffic. In: Nossal H, Vogel H (eds) The pathobiology of the endothelial cell. Academic, New York, pp 409–424

Butcher EC, Rouse RV, Coffman RL, Nottenburg C, Hardy RH, Weissman IL (1982c) Surface phenotype of Peyer's patch germinal center B cells: implications for the role of germinal centers in B cell differentiation. J Immunol 129:2698

Butcher EC, Lewinsohn D, Duijvestijn A, Bargatze R, Wu N, Jalkanen S (1986) Interactions between endothelial cells and leukocytes. J Cell Biochem 30:121–131

Cahill RNP, Frost H, Trnka Z (1976) The effects of antigen on the migration of recirculating lymphocytes through single lymph nodes. J Exp Med 143:870

Cahill RNP, Poskitt DC, Frost H, Trnka A (1977) Two distinct pools of recirculating T lymphocytes: migratory characteristics of nodal and intestinal T lymphocytes. J Exp Med 145:420–428

Cahill RNP, Heron I, Poskitt DC, Trnka A (1980) Lymphocyte recirculation in the sheep fetus. In: Blood cells and vessel walls: functional interactions. Excerpta Medica, Amsterdam, pp 145–166 (Ciba Foundation symposium no 71)

Carey GD, Chin Y-H, Woodruff JJ (1981) Lymphocyte recognition of lymph node high endothelium. III. Enhancement by a component of thoracic duct lymph. J Immunol 127:976–979

Cavender DE, Haskard DO, Joseph B, Ziff M (1986) Interleukin 1 increases the binding of human B and T lymphocytes to endothelial cell monolayers. J Immunol 136:203–207

Chin W, Hay JB (1980) A comparison of lymphocyte migration through intestinal lymph nodes, subcutaneous lymph nodes, and chronic inflammatory sites of sheep. Gastroenterology 79:1231–1242

Chin Y-H, Rasmussen RA, Woodruff JJ, Easton TG (1986) A monoclonal anti-HEBF antibody with specific for lymphocyte surface molecules mediating adhesion to Peyer's patch high endothelium of the rat. J Immunol 136:1–5

Coico RF, Bhogal BS, Thorbecke GJ (1983) Relationship of germinal centers in lymphoid tissue to immunologic memory. VI. Transfer of B cell memory with lymph node cells fractionated according to their receptors for peanut agglutinin. J Immunol 131:2254–2257

Dailey MO, Fathman CG, Butcher EC, Pillemer E, Weissman I (1982) Abnormal migration of T-lymphocyte clones. J Immunol 128:2134–2136

Dailey MO, Gallatin WM, Weissman IL, Butcher EC (1983) Surface phenotype and migration properties of activated lymphocytes and T cell clones. In: Parker JW, O'Brien RL (eds) Intercellular communication in leukocyte function. Wiley, New York, pp 641–644

Duijvestijn AD, Schreiber A, Butcher EC (1986) Interferon-gamma induces an antigen specific for endothelium involved in lymphocyte homing. Proc Nat Acad Sci USA, in press

Elson CO, Heck JA, Strober W (1979) T cell regulation of murine IgA synthesis. J Exp Med 149:632

Fink PJ, Gallatin WM, Reichert RA, Butcher EC and Weissman IL (1985) Homing receptor-bearing thymocytes, an immunocompetent cortical subpopulation. Nature 313:233–235

Ford WL (1975) Lymphocyte migration and immune responses. Prog Allergy 19:38

Ford WL, Smith ME, Andrews P (1978) Possible clues to the mechanism underlying the selective migration of lymphocytes from the blood. In: Curtis ASG (ed) Cell-cell recognition. Cambridge University Press, Cambridge, pp 359–392

Fossum S, Smith ME, Ford WL (1983) The recirculation of T- and B-lymphocytes in the athymic, nude rat. Scand J Immunol 17:551–557

Gallatin WM, Weissman, Butcher EC (1983) A cell surface molecule involved in organ-specific homing of lymphocytes. Nature 304:30–34

Gowans JL, Knight EJ (1964) The route of recirculation of lymphocytes in the rat. Proc R Soc Lond [Biol] 159:257

Gray D, MacLennan ICM, Bazin H, Khan M (1982) Migrant $\mu^+\delta^+$ and $\mu^+\delta$ static B lymphocyte subsets. Eur J Immunol 12:564

Gresser I, Guy-Grand D, Maury C, Maunoury MT (1981) Interferon induces peripheral lymphadenopathy. J Immunol 127:1569–1575

Griscelli C, Vassalli P, McCluskey RT (1969) The distribution of large dividing lymph node cells in syngeneic recipient rats after intravenous injection. J Exp Med 130:1427–1451

Guy-Grand D, Vassalli P (1982) Nature and function of gut granulated T-lymphocytes. In: Strober W, Hanson LA, Sell KW (eds) Recent advances in mucosal immunity. Raven, New York, p 301

Guy-Grand D, Griscelli C, Vassalli P (1974) The gut-associated lymphoid system: nature and properties of the large dividing cells. Eur J Immunol 4:435–443

Guy-Grand D, Griscelli C, Vassalli P (1978) The mouse gut T-lymphocyte, a novel type of T cell. Nature, origin and traffic in mice in normal and graft-versus-host conditions. J Exp Med 148:1661–1667

Hall JG, Parry DM, Smith ME (1972) The distribution and differentiation of lymph-borne immunoblasts after intravenous injection into syngeneic recipients. Cell Tissue Kinet 5:269

Hall JG, Hopkins J, Orlans E (1979) Studies on the lymphocytes of sheep. III. Destination of lymph-borne immunoblasts in relation to their tissue of origin. Eur J Immunol 7:30–37

Halstead TE, Hall JG (1972) The homing of lympho-borne immunoblasts to the small gut in neonatal rats. Transplantation 14:339

Hamann A, Jablonski-Westrich D, Raedler A, Thiele HG (1984) Lymphocytes express specific antigen-independent contact interaction sites upon activation. Cell Immunol 86:14–32

Hamann A, Jablonski-Westrich D, Scholz K-U, Duijvestijn A, Butcher EC, Thiele HG (1986) Activation results in complex alterations in homing receptor expression and migratory properties of lymphocytes. (To be published)

Hay JB, Johnston MG, Vadas P, Chin W, Issekutz T, Movat HZ (1980) Relationship between changes in blood flow and lymphocyte migration induced by antigen. Monogr Allergy 16:112–125

Hendriks HR, Estermans IL (1983) Disappearance and reappearance of high endothelial venules and immigrating lymphocytes in lymph nodes deprived of afferent lymphatic vessels: a possible regulatory role of macrophages in lymphocyte migration. Eur J Immunol 13:663–669

Hopkins J, McConnell I, Pearson JD (1981) Lymphocyte traffic through antigen-stimulated lymph nodes. II. Role of prostaglandin E2 as a mediator of cell shutdown. Immunology 42:225–231

Husband AJ (1982) Kinetics of extravasation and redistribution of IgA-specific antibody-containing cells in the intestine. J Immunol 128:1355–1359

Husband AJ, Gowans JL (1978) The origin and antigen-dependent distribution of IgA-containing cells in the intestine. J Exp Med 148:1146–1160

Husband AJ, Monie HG, Gowans JL (1977) The natural history of the cells producing IgA in the gut. In: Immunology of the gut. Excerpta Medica, Amsterdam (Ciba Foundation symposium no 46)

Jalkanen ST, Butcher EC (1985) *In vitro* analysis of the homing properties of human lymphocytes: developmental regulation of functional receptors for high endothelial venules. Blood 66:577–582

Jalkanen S, Bargatze R, Herron L, Butcher EC (1986a) A lymphoid cell surface glycoprotein involved in endothelial cell recognition and lymphocyte homing in man. Eur J Immunol, in press

Jalkanen S, Steere A, Fox R, Butcher EC (1986b) A distinct endothelial cell recognition system controlling lymphocyte traffic into inflamed synovium. Science, in press

Jalkanen S, Bargatze R, Butcher EC (1986c) A gp 90 receptor class mediates several organ-specific lymphocyte-endothelial recognition events in man. (To be published)

Kieda C, Monsigny M (1983) The adhesion of mouse spleen cells to lymph node venules involves endogenous membrane lectins (sugar receptors) of the lymphocytes. In: Parker JW, O'Brien RL (eds) Intercellular communication in leukocyte function. Wiley, New York, p 649

Klaus GGB, Kunkl A (1981) The role of germinal centres in the generation of immunological memory. In: Microenvironments in haemopoietic and lymphoid differentiation. Excerpta Medica, Amsterdam, pp 265–280

Korngold R, Blank KJ, Murasko DM (1983) Effect of interferon on thoracic duct lymphocyte output: induction with either poly I:poly C or vaccinia virus. J Immunol 130:2236–2240

Kraal G, Weissman IL, Butcher EC (1982) Germinal center B cells: antigen specificity and changes in heavy-chain isotype expression. Nature 298:377

Kraal G, Weissman IL, Butcher EC (1983) Differences in in vivo distribution and homing of T cell subsets to mucosal vs. non-mucosal lymphoid organs. J Immunol 130:1097–1102

Kraal G, Weissman IL, Butcher EC (1985) Germinal center cells: antigen specificity, heavy chain class expression and evidence of memory. Adv Exp Med Biol 186:145–151

Kraal G, Hardy RR, Gallatin WM, Weissman IL, Butcher EC (1986) Antigen-induced changes in B cell subsets in lymph nodes: analysis by dual fluorescence flow cytofluorometry. (To be published)

Lamm ME (1976) Cellular aspects of immunoglobulin A. Adv Immunol 22:223–290

Lewinsohn D, Bargatze R, Butcher EC (1986) A common endothelial cell recognition system shared by neutrophils, lymphocytes, and other leukocytes. (To be published)

Marchesi VT, Gowans JL (1964) The migration of lymphocytes through the endothelium of venules in lymph nodes. Proc R Soc Lond [Biol] 159:283

McConnell I, Hopkins J (1981) Lymphocyte traffic through antigen-stimulated lymph nodes. I. Complement activation within lymph nodes initiates cell shutdown. Immunology 42:217–223

McCullagh P (1980) Unresponsiveness of recirculating lymphocytes after antigenic challenge. Monogr Allergy 16:143–156

McDermott MR, Bienenstock J (1979) Evidence for a common mucosal immunologic system. I. Migration of B immunoblasts into intestinal, respiratory and genital tissues. J Immunol 122:1892

McWilliams M, Phillips-Quagliata JM, Lamm ME (1975) Characteristics of mesenteric lymph node cells homing to gut-associated lymphoid tissue in syngeneic mice. J Immunol 115:54–58

McWilliams M, Phillips-Quagliata JM, Lamm ME (1977) Mesenteric lymph node B lymphoblasts which home to the small intestine are pre-committed to IgA synthesis. J Exp Med 145:866–875

Moore TC, Lachmann PJ (1982) Cyclic AMP reduces and cyclic GMP increases the traffic of lymphocytes through peripheral lymph nodes of sheep in vivo. Immunology 47:423–428

Navarro RF, Jalkanen ST, Hsu M, Goronzy J, Weyand C, Fathman G, Clayberger C, Krensky J, Butcher EC (1985) Human T cell clones express functional homing receptors required for normal lymphocyte trafficking. J Exp Med 162:1075–1080

Ottaway CA, Parrott DMV (1979) Regional blood flow and its relationship to lymphocyte and lymphoblast traffic during a primary immune reaction. J Exp Med 150:218

Parrot DMV, Ferguson A (1974) Selective migration of lymphocytes within the mouse small intestine. Immunology 26:571

Rannie GH, Donald KJ (1977) Estimation of the migration of thoracic duct lymphocytes to nonlymphoid tissues. Cell Tissue Kinet 10:523

Rasmussen RA, Chin YH, Woodruff JJ, Easton TG (1985) Lymphocyte recognition of lymph node high endothelium. VII. Cell surface proteins involved in adhesion defined by monoclonal anti-HEBFLN (A. 11) antibody. J Immunol 135:19–24

Reichert RA, Weissman IL, Butcher EC (1983) Germinal center cells lack homing receptors necessary for normal lymphocyte recirculation. J Exp Med 157:813–827

Reichert RA, Gallatin WM, Butcher EC, Weissman IL (1984) A homing receptor bearing cortical thymocyte subpopulation: implications for thymus cell migration and the nature of cortisone-resistant thymocytes. Cell 38:89–99

Reichert RA, Gallatin WM, Weissman IL, Butcher EC (1986a) Phenotypic analysis of thymocytes that express homing receptors for peripheral lymph nodes. J Immunol (to be published)

Reichert RA, Jerabek L, Gallatin WM, Butcher EC, Weissman IL (1986b) Ontogeny of lymphocyte homing receptor expression in the mouse thymus. J Immunol (to be published)

Reynolds J, Heron I, Dudler L, Trnka Z (1982) T-cell recirculation in the sheep: Migratory properties of cells from lymph nodes. Immunology 47:415–421

Rose ML, Parrott DMV, Bruce RG (1976) Migration of lymphoblasts to the small intestine. II. Divergent migration of mesenteric and peripheral immunoblasts to site of inflammation in the mouse. Cell Immunol 27:36–46

Rose ML, Parrott DMV, Bruce RG (1978) Accumulation of immunoblasts in extravascular tissues including mammary gland, peritoneal cavity, gut and skin. Immunology 35:415

Rose ML, Birbeck MSC, Wallis VJ, Forrester JA, Davies AJS (1980) Peanut lectin binding properties of germinal centers in mouse lymphoid tissue. Nature 284:364–366

Rosen SD, Singer MS, Yednock TA, Stoolman LM (1985) Involvement of sialic acid on endothelial cells in organ-specific lymphocyte recirculation. Science 228:1005–1007

Rowley DA, Gowans JL, Atkins RC, Ford RC, Smith WL (1972) The specific selection of recirculating lymphocytes by antigen in normal and preimmunized rats. J Exp Med 136:499

Schmitz M, Nunez D, Butcher EC (1986) Selective recognition of mucosal endothelium by gut intraepithelial leukocytes. (To be published)

Scollay R, Hopkins J, Hall J (1976) Possible role of surface Ig in non-random recirculation of small lymphocytes. Nature 260:528–529

Scollay RG, Butcher EC, Weissman IL (1980) Thymus cell migration: quantitative aspects of cellular traffic from the thymus to the periphery in mice. Eur J Immunol 10:210–218

Smith JG, McIntosh GH, Morris B (1970) The traffic of cells through tissues: a study of peripheral lymph in sheep. J Anat 107:87

Smith ME, Martin AF, Ford WL (1980) Migration of lymphoblasts in the rat: preferential localization of DNA-synthesizing lymphocytes in particular lymph nodes and other sites. Monogr Allergy 16:203–232

Spangrude GJ, Braaten BA, Daynes RA (1984) Molecular mechanisms of lymphocyte extravasation. I. Studies of two selective inhibitors of lymphocyte recirculation. J Immunol 132:354–362

Sprent J (1973) Circulating T and B lymphocytes of the mouse. I. Migratory properties. Cell Immunol 7:10

Sprent J (1977) Recirculating lymphocytes. In: Marchalonis JJ (ed) The lymphocyte: structure and function, chap 2. Dekker, New York

Sprent J (1980) Antigen-induced selective sequestration of T lymphocytes: role of the major histocompatibility complex. Monogr Allergy 16:233–244

Stamper HB Jr, Woodroff JJ (1976) Lymphocyte homing into lymph nodes: In vitro demonstration of the selective affinity of recirculating lymphocytes for high endothelial venules. J Exp Med 144:828–833

Stevens SK, Weissman IL, Butcher EC (1982) Differences in the migration of B and T lymphocytes: organ-selectivity and the role of lymphocyte-endothelial cell recognition. J Immunol 128:844–851

Stoolman LM, Rosen SD, Tenforde T (1983) Phosphorylated carbohydrates inhibit an adhesive interaction implicated in lymphocyte recirculation. Abstract 339. J Cell Biol 97:88a

Stoolman LM, Tenforde TS, Rosen SD (1984) Phosphomannosyl receptors may participate in the adhesive interaction between lymphocytes and high endothelial venules. J Cell Biol 99:1535–1540

Streilein JW (1978) Lymphocyte traffic, T cell malignancies and the skin. J Invest Dermatol 71:167–171

Strober S, Dilley J (1973) Maturation of B lymphocytes in the rat. I. Migration pattern, tissue distribution and turnover rate of unprimed and primed B lymphocytes involved in the antidinitrophenyl response. J Exp Med 138:1331

Thorbecke GJ, Romano TJ, Lerman SP (1974) In: Brent L, Holborow J (eds) Progress of immunology II. vol 3. North-Holland, Amsterdam, pp 25–34

Trnka Z, Cahill RNP (1980) Aspects of the immune response in single lymph nodes. Monogr Allergy 16:245–259

van Dinther-Janssen AC, van Maarsseveen AC, DeGroot J (1983) Comparative migration of T- and B-lymphocyte subpopulations into skin inflammatory sites. Immunology 48:519–527

Woodruff JJ, Katz IM, Lucas LE, Stamper HB Jr (1977) An in vitro model of lymphocyte homing. II. Membrane and cytoplasmic events involved in lymphocyte homing. J Immunol 119:1603

Yednock TA, Butcher EC, Stoolman LM, Rosen SD (1986) Lymphocyte-homing receptors: relationship between the MEL-14 antigen and a carbohydrate-binding receptor. (To be published)